The Emergence of
Life on Earth

THE EMERGENCE OF LIFE ON EARTH

A Historical and Scientific Overview

IRIS FRY

RUTGERS UNIVERSITY PRESS
New Brunswick, New Jersey, and London

Library of Congress Cataloging-in-Publication Data

Fry, Iris.
    The emergence of life on Earth : a historical and scientific
overview / Iris Fry.
        p. cm.
    Includes bibliographical references (p.).
    ISBN 0-8135-2739-2 (cloth : alk. paper). — ISBN 0-8135-2740-6
(paper : alk paper)
        1. Life—Origin. I. Title.
    QH325.F78 2000
    576.8'3—dc21   99-23153
                                                             CIP

British Cataloging-in-Publication data for this book is available from the British
Library

Manufactured in the United States of America

To my family, with love

Contents

Acknowledgments

Parts of this book are based on my book, in Hebrew, *The Origin of Life—Mystery or Scientific Problem?* published in 1997 by the Broadcast University Library of the Ministry of Defense Press, Israel. I wish to thank the Israeli publisher for granting me permission to use the Hebrew material, which served as stimulus for the writing of this volume.

I am indebted to my friend Marcella Benditt, whose profound command of the English language and remarkable editorial skills helped to shape the form of this book. Her enthusiasm for the book was invaluable.

Adam Rabinowitz's skillful preparation of some of the illustrations is gratefully acknowledged.

Thanks are due to Helen Hsu, my editor at Rutgers University Press, for her help and encouragement.

Last but not least, thanks to my friend and husband, Mickey, without whom this book would not have been possible.

The Emergence of
Life on Earth

INTRODUCTION
THE ORIGIN OF LIFE—
A SUBJECT WITH A PAST
AND A FUTURE

Is there life on other worlds? Can we search for it within our solar system and beyond? How similar is extraterrestrial life to life on Earth, and are we bound to find life only as we know it? These questions, until not long ago confined to the pages of speculative essays and to science fiction books and tabloid headlines, are now the legitimate subject matter of extensive scientific research. Recent discoveries of extrasolar planets and the tentative indications of life on other celestial bodies have drawn public attention also to the question of the emergence of life on Earth. This interest is but the latest stage in a long historical drama. For hundreds of years the origin of life was thought to be well understood and did not pose any problem to naturalists and lay persons alike. Interestingly, doubts began to rise with the growth of biological knowledge, and the question became a complete mystery, especially at the end of the nineteenth and the beginning of the twentieth century. It is only in the last decades that the emergence of life on Earth has entered the realm of investigation as a scientific problem.

This book will trace these developments, including a detailed discussion of current lines of research devoted to the origin of life. Unlike most other writers who have addressed this topic, I examine it from a combined scientific, historical, and philosophical perspective. The first part of the book is historical, offering an overview of the development of the main ideas about the origin

1

of life from antiquity to the present century. The second part, which is much more detailed, examines the major prevailing scientific theories of the origin of life. Since this book is intended for readers interested in science, and since in many cases there is a "division of interests" between science-oriented and history-oriented readers, the question, Why bother at all with history? has to be addressed. Why, indeed, not discard the first, historical part, get straight to the point, and discuss what we know, and do not yet know, about the emergence of life on Earth? Though an impatient reader can jump right into the twentieth century in chapter 6, there are several important reasons for a historical examination of the subject. First, as I hope to show, it involves an interesting story, full of anecdotes and what to our eyes are rather strange and surprising beliefs. Second, the tendency among some scientists to focus on the "here and now" and to ignore older ideas goes against the grain of science itself. Scientists fully appreciate the role of historical developments in the realms of physical and biological reality. However, they sometimes fail to appreciate that our concepts likewise do not materialize out of thin air, but owe their origin and character to past developments, both successes and failures. The ties that connect present ideas on the emergence of life to older traditions will be described in the book.

In addition to the previous reasons justifying a historical examination, even a brief presentation of the history of ideas about the origin of life can serve to illustrate the many factors that shape our evaluation of nature. Attitudes toward the problem of the emergence of life have always been determined by observations and experiments and also by philosophical, religious, and political factors. My historical analysis demonstrates that the examined facts themselves, as well as common sense, were conditioned by many nonempirical presuppositions. Hence, not only did different beliefs depend each on its own conceptual framework, but the overturn of these beliefs did not result necessarily from empirical refutation. The willingness to consider empirical results that seemed to refute a previously held theory depended also on extra-scientific factors. The lesson learned from examining past ideas should help enlighten us about the role of philosophical and religious preconceptions in the controversies currently surrounding the origin-of-life issue. It might also lead us to reexamine the role of cultural and social factors in the working of science in general.

Throughout human history, naturalists, philosophers, and lay persons alike have been preoccupied with and fascinated by the subject of the emergence of life. It might come as a surprise to some readers to learn that for most of history, the origin-of-life question was phrased very differently from the way it is formulated today, not focusing at all on the ancient Earth. The general belief was that many forms of life kept originating constantly in a process called "spontaneous generation," not from any parent, but directly from inorganic and organic matter. The belief in spontaneous generation was so widespread and common as to be shared in different eras by people of very different philosophical inclinations. It was backed, throughout the ages, by both materialistic and religious arguments that stemmed from people's differing perceptions of the relationship between inanimate matter and life. Historically, the doctrine of spontaneous generation died a slow death, especially as far as the spontaneous generation of microorganisms was concerned. It was finally abandoned toward the end of the nineteenth century, following several scientific and philosophical developments that will be discussed in this book, most notably the famous experiments of the French scientist Louis Pasteur.

Alongside the doctrine of spontaneous generation, and sometimes in conflict with it, there reigned for many centuries the belief in the separate divine creation of the biological species and their unchanging existence since creation. It was Charles Darwin's theory of evolution, published in 1859 in *The Origin of Species*, that challenged the belief in separate creation. Darwin made a very convincing case, on the basis of a vast amount of data, for the claim that the biological species developed one from another in the course of the Earth's history. The revolutionary significance of his theory stemmed mainly from the mechanism of natural selection, which he offered as the major explanation for the facts of evolution. The revolutionary philosophical implications of natural selection lay in the option, now made possible, to replace the former explanations of living nature based on divine design with explanations based on natural, material processes.

A basic postulate of the theory of evolution is that all organisms existing today evolved from a common ancestor. Comparative studies of the key macromolecules in many different species indeed point to a convergence toward such a common root at the base of the "tree of life." It is this postulate that provides a reasonable explanation of the intricacy and of the multilevel similarities among the tree's branches. Darwin spoke about "a few forms or . . . one" from which all the living world around us evolved. There were attempts to reconcile a belief in the divine creation of these original life-forms with the theory of evolution; Darwin himself mentioned this possibility. Logically, however, it is evident that a consistent evolutionary theory should address the question of the material origin of the first living systems on Earth. Historically, this was how Darwin's theory was conceived by most of its adherents and by its enemies. And yet there is no discussion of the origin of life in *The Origin of Species*. As becomes clear from some of Darwin's letters, he thought it futile, considering the state of contemporary scientific knowledge, even to raise the question of the development of life from inanimate matter. In the long run, it was the rise of a new scientific and philosophical worldview, fostered by Darwin's theory, that was responsible for the reformulation of the origin-of-life question in evolutionary terms during the first decades of the twentieth century. In this new form the subject was finally divorced from doctrines of both spontaneous generation and divine creation, and instead conceived in terms of specific, gradual processes of chemical evolution of the first living systems on the primordial Earth.

Following complex philosophical and scientific developments, a research field devoted to the theoretical and empirical investigation of the emergence of life was established in the early 1950s. The basic evolutionary premise of this research is the acknowledgment of both the continuity and the radical novelty involved in the emergence of life from inanimate matter. Cosmic, chemical, and biological processes are part of a historical continuum with no unbridgeable gaps between inorganic, organic, and biological matter. Yet the newly evolved living systems were characterized by unique properties that did not exist before. To account scientifically for both continuity and change in this passage, there is need for cooperation among the disciplines of physics, chemistry, and biology. Origin-of-life research is, by definition, an interdisciplinary field to which, among others, astrophysicists, planetary scientists, geologists, physical, organic, and biological chemists, evolutionary biologists, and molecular biologists all contribute. This interdisciplinary field is fraught with controversy and uncertainties: When in Earth's history did life emerge, and where did the process take place? What were the conditions on Earth during this period? Did life emerge on Earth at all, or was it brought to Earth from outer space? Were the organic building blocks necessary for life's emergence synthesized here, or were they "imported" from outside? Which was the first living system to emerge? Was it a primitive cell-like structure, a set of "primitive enzymes," or a primitive replicating molecule? As will be pointed out later, underlying some of the factual, empirical debates is a basic division of principle between the "genetic approach," which claims that the first to emerge was a genetic system, and the "metabolic approach," according to which the first living entity was an autocatalytic metabolic cycle, presumably composed of protein-like molecules. These two approaches are the continuation of two distinct traditions, both of which developed at the beginning of the twentieth century.

Evolutionary biologists agree that the last common ancestor from which all organisms descended was already a complex cellular system containing many of the basic biochemical features found in all living creatures. Current origin-of-life theories concentrate on mechanisms of chemical evolution and the beginning of biological evolution thought to have taken place prior to the evolution of the common ancestor. These processes might have led to the emergence of primitive systems that eventually gave rise to the common ancestor. Since it is impossible to reconstruct the actual many-stepped path taken by emerging life, the theoretical emphasis is on the physicochemical principles and possible mechanisms that could have led to life. At the same time, however, astronomical and geophysical investigations increasingly provide information on the actual conditions believed to have prevailed on the ancient Earth.

The notion often presented in the popular literature that science has already solved the origin-of-life question has to be seriously qualified. This book will demonstrate that the hurdles impeding a solution are still numerous. Nevertheless, both the optimists among the active researchers in the field, who foresee a relatively quick resolution, and the pessimists, who don't, are convinced that life emerged in the past through natural, material processes with no supernatural intervention. Acknowledging the problems and difficulties that the origin-of-life field is facing, the common conviction is that science will eventually unravel the working of these processes. Based on our growing understanding of the characteristics of living systems and on the accumulated knowledge of the physical and chemical conditions on Earth and in the solar system billions of years ago, and within the framework of known natural laws, the origin of life can be approached today not as a mystery but as a scientific problem.

As part of my discussion of the current lines of research on the origin of life, I will call attention to recent indications of the possibility of life existing on various celestial bodies near and far. The findings published since 1996 by NASA scientists of organic compounds in a meteorite rock that, in all probability, reached Earth from Mars, as well as the tentative, highly disputed, claims that this rock contains the remains of fossilized microorganisms that might have originated on Mars, have created quite a stir among lay people and scientists. Though most scientists do not believe now that this particular rock contains evidence of past Martian life, studies of this and other Martian meteorites continue. Future missions to Mars, scheduled to take place in the next few years, are planned to bring back rock samples that might contain evidence of past or present biological activity. Research is also focused on chemical processes that might have led to life, and more boldly, on the possibility of life itself on several other bodies in the solar system. One of Jupiter's moons, Europa, might have a warm ocean below its icy external layer, possibly containing organic compounds and maybe even life. One of Saturn's moons, Titan, manifests a highly complex organic chemistry in its atmosphere. The search for a possible abode for life outside the solar system largely coincides with the search for planetary systems, with the hope of finding Earth-like planets that might harbor conditions for life. Until 1995, the assumption that the solar system is just one example among many was solely based on the Copernican reasoning that our planet is not unique. The discovery since then of quite a few planets around sun-like stars lends factual support to the claim that the formation of planetary systems in the universe is a widespread phenomenon. Furthermore, extending the Copernican conception, many scientists hold that the occurrence of the necessary, if not sufficient, conditions for the emergence of life is common in the universe.

Interestingly, the search for life in outer space—called alternatively exobiology, bioastronomy, or astrobiology—has recently received a very serious boost from new discoveries about life on Earth. Terrestrial microorganisms, using various chemical substances as sources of energy, have been discovered thriving under conditions previously thought impossible for the existence of life: in hot springs at the bottom of the ocean, in the coldest areas of Antarctica, and deep under basalt rocks. There are strong indications that these "extremophiles" evolved very early in the history of life and are similar in their molecular composition to the hypothetical last common ancestor. The speculation that life might have emerged deep in the ocean under Europa's ice cap or in underground springs on Mars is clearly inspired by these findings on the home front. Both research in astrobiology and the discovery of previously unknown microorganisms in harsh environments on Earth focus the interest of astronomers and geophysicists on the question of possible settings for the origin of life. This "setting-focused" approach, by studying closely the environmental conditions under which life might have emerged, helps to define the limits of a historically feasible origin-of-life theory.

As we shall see, indications that life on Earth arose very rapidly, almost "immediately" when geological conditions allowed, tend to strengthen the belief in the frequent emergence of life on other planets, provided that appropriate conditions prevailed. We should be aware, though, that the issue of the probability of the emergence of life on any planet involves two categorically distinct questions. The first deals with the frequency of life-bearing, or "biogenic," conditions on celestial bodies, and is to be decided on the basis of astronomical data. The second considers the chances for the emergence of life, on Earth and elsewhere, once the appropriate physicochemical conditions are provided. The answer to this question involves not only empirical knowledge but also philosophical notions. It is the claim of some biologists—fewer today than in the past—that due to the enormous complexity of even the most primitive living system, chances of its emergence are extremely small. Some of these scientists view the origin of life as a rare "happy accident," as "almost a miracle." Relying on a similar argument but drawing from it very different conclusions, creationists deny the natural emergence of life and uphold the necessity of a divine intervention. This book will provide scientific and philosophical arguments denying such claims. In agreement with most researchers in the origin-of-life field today, it is my contention that within the realistic confines of space and time of our universe, the emergence of life could not have been the result of chance, nor was it a miracle fostered by a supernatural power. Rather it involved the working of physical and chemical mechanisms responsible for the self-organization of matter into living systems. Such mechanisms, given the appropriate environmental conditions, could have produced similar results elsewhere in the universe.

This position, though based on several empirical considerations, is first and foremost an expression of a philosophical worldview that denies the separation of living systems and inanimate matter into two unbridgeable categories and provides an implicit guideline for origin-of-life research. Within this framework, different theories are formulated, and possible specific mechanisms are devised. Is this naturalistic, evolutionary worldview, as is sometimes claimed by the creationists, simply a dogma, a system of beliefs, no better or worse than a religious dogma? Not so. First, the historical development of this worldview is based on the accumulation of vast amounts of empirical data relating to all facets of nature. The evolutionary conception is backed by a well-established body of knowledge embracing the origin and evolution of life as well as the whole domain of the natural sciences. Second, taking into account the epistemological and social constraints within which science works and the inevitable dependence of its empirical claims on philosophical, theoretical, and cultural presuppositions, it still differs from a religious system of beliefs.

The hypotheses put forward by science, including those on the origin of life, have to withstand the criteria of criticism and testing. Challenged by the encounter with recalcitrant data, these hypotheses are very often found to be untenable and in need of replacement. As will be explained in this book, the various hypotheses currently advanced by origin-of-life researchers, while making a significant contribution to our understanding of the subject, do not yet seem to suggest a convincing solution to the problem of the emergence of life. But no origin-of-life scientist considers that a reason to forsake the naturalistic worldview for the creationist one. Despite the many conflicts and divisions marking this field of research, its common philosophical premise is that the emergence of life did not require the helping hand of a "designer," nor did it involve an extraordinary and rare leap across a great divide.

1

"FROM THE DUST OF THE GROUND"

Throughout most of human history—in ancient times, during the Middle Ages, and even in the Modern Age—a general belief held that living organisms can arise not only from parents, but also spontaneously: from inorganic and organic matter, independent of any parent. It was widely claimed that plants, worms, insects, and, under certain circumstances, even fish, frogs, and birds can be generated from soil, tree bark, slime, excrement, and decaying plant and animal matter, under the influence of heat and moisture. This common conviction was based on everyday observations of the sudden appearance of all sorts of living creatures, such as insects in putrefying meat, worms in muddy areas, and frogs and mice in moist earth.

Among the supporters of spontaneous generation throughout history have been people of very different beliefs and with philosophical inclinations. It is clear that this doctrine was supported by different lines of reasoning in different eras. In ancient Greece and again in seventeenth-century Europe, a materialistic explanation was given, according to which there is no real distinction between inanimate and animate matter and hence life can be easily generated from matter. An opposing view, which persisted for many centuries, was that the sudden generation of living creatures from matter, independent of parents, was due to direct or indirect divine intervention.

Stories about the recurrent generation of life out of inanimate matter are to be found in all ancient mythologies, usually portraying some divine power as the responsible agent (Oparin 1968:9–11; Abel 1973:16–18). According to the biblical story in Genesis 2, God created man, as well as the beasts of the field and the birds of the air, from the dust of the ground. Another Genesis version, in chapter 1, refers to a divine decree imposed on the land and the water, out of which plants and animals were formed. Ancient Chinese, Babylonian, and Indian scriptures describe the generation of insects from sweat and manure and the appearance of worms and other creatures from mud in canals and rivers. The mud of the Nile, under the influence of the sun, was considered to be especially "fertile" for the formation of snakes and crocodiles (Diodorus 1960). This belief is reflected, for example, in Shakespeare's *Antony and Cleopatra* when Lepidus tells Antony, "Your serpent of Egypt is bred now of your mud by the operation of your sun: so is your crocodile" (act 2, scene 7).

THE GREEK PHILOSOPHERS OF NATURE—THE FIRST THEORIES OF MATTER

The rise of a scientific view of nature, which occurred in Greece during the seventh and sixth centuries B.C.E., was expressed in the move from a mythological interpretation of experiences and observations to an exploration of natural causes for natural phenomena (Burnet 1930; Sambursky 1960). During this philosophical transition, Greek thinkers sought to explain their belief in spontaneous generation by various theories about the constitution of matter and by their general philosophy of nature. According to these theories of matter, all objects were believed to be constituted of one basic stuff, and within this theoretical framework, living organisms were thought to be spontaneously generated from this material (Toulmin and Goodfield 1982:48–52). The philosopher Thales assumed that everything is made of water. Thales' pupil and successor Anaximander claimed that the basic element, the "first principle," was infinite and imperceptible; thus he called it "the Apeiron"—the boundless. Anaximander's pupil Anaximenes spoke of the air as the basic element, whereas Heraclitus adopted fire as the basic substance (Kirk and Raven 1957).

In the fifth century B.C.E., the Sicilian philosopher Empedocles hypothesized that nature is built from not one basic stuff but four eternal elements—water, earth, fire, and air—which he termed "roots." Under the influence of two primary natural forces—love and strife—the four elements repeatedly combine with each other in different proportions (thus creating all the composite objects, including living creatures) and repeatedly fall apart. The Empedoclean conception of the four elements dominated Western cosmology and medical thought for hundreds of years, until the Renaissance. Empedocles, like other philosophers at different times, regarded the history of the cosmos as cyclical, undergoing periodic rounds of creation and decay governed by the power of love and harmony and the power of strife and separation. In each of these cycles, out of the chance combinations of the four elements, various living creatures and parts of such creatures are formed. From time to time, these accidentally formed creatures are monstrous, such as a man bearing an ox's head, or an ox with a human head, creatures having double faces and torsos, and odd combinations of male and female organs. Few of these creatures survive: only those whose parts happen to fit together and who can nourish themselves and breed (Lambridis 1976:92–105).

The ideas of Empedocles are sometimes compared to those of Darwin (Lambridis 1976:101–104; Abel 1973:39–40). In particular, attention is drawn to "natural selection" operating on the various living combinations, which results in the extinction of the unfit monsters. In distinction to Aristotle, to be shortly discussed, who saw the basic species—the basic forms of the living world—as eternal, Empedocles described the generation and extinction of different species. He also emphasized the role of chance in natural processes. These features are frequently hailed as an expression of the modern spirit in his thought. It is important, however, to be on guard against the tendency in some historical analyses to dislocate concepts from their historical contexts and to associate them with ideas of different times and places. Despite similar elements, Darwinian thought of the nineteenth century is completely foreign to Empedocles and his time. The four basic elements are eternal, and the material universe goes through cycles of eternal reconstruction. Unlike Darwin, Empedocles does not speak of the development or descent of species out of previous species (Mayr 1982:301–303). His "theory of evolution" is in fact an intricate description of the spontaneous generation of living creatures out of the four elements (Abel 1973:40). It should be noted that unlike Aristotle, Empedocles does not dwell on the unique nature of living organisms. In his view, all natural objects, including organisms, are the outcome of "mixing and separating" of the four elements. He does not acknowledge the significance of the organic "form," or in our terms, of biological organization. In radical distinction to Aristotle, Empedocles ignores the complex, functional, self-preserving organization that reconstructs itself, not by chance but repeatedly and according to a specific order, during the life of each organism and in the passage from generation to generation.

During the fifth, fourth and third centuries B.C.E., Greek philosophy of nature took a new direction, which had a protracted influence on Western thought. The Greek atomists Leucippus, Democritus, and Epicurus developed a materialistic philosophical system, explaining natural phenomena, including the human soul and the development of human society, in terms of causal materialistic processes. They rejected the use of teleological explanations and reliance on the intervention of the gods in natural and human affairs (Toulmin and Goodfield 1982:56–72). The atomists postulated the existence of an infinite number of tiny, invisible, indivisible, corporeal bodies—atoms—as the basic constituents of the world. Atoms are in constant motion within infinite empty space, and it is as a result of the random collisions among them that all material objects, including organisms, are formed. All atoms are made of the same basic stuff, but differ in size and shape. When atoms corresponding in form meet randomly, they combine and gradually form the more complex bodies. The atomists' ideas were forcefully presented by the Roman poet Lucretius Carus (99–55 B.C.E.) in his poem *On the Nature of Things* (*De rerum natura*) (Lucretius 1956). Lucretius also describes the formation of the earth from the combination of atoms, and the formation of plants and animals from the earth. In the past and even today, says Lucretius, "mother earth" gives life to various creatures (Lucretius 1956:221).

Like Empedocles, Lucretius described the spontaneous generation of unfit creatures and of monsters of all kinds, who could not grow and reproduce their kind and therefore did not survive (Lucretius 1956:223–224). Again, despite the apparent resemblance to modern materialistic ideas, the atomists did not formulate a theory of evolution. They did not postulate "descent with modification" of one species from another, but rather offered a sophisticated theory that described the spontaneous generation of different organisms in various stages of the development of the earth.

Discussions of the atomists' philosophy, in ancient Greece and in later periods, deal with the role of chance in natural processes. But it is important to understand that in addition to their emphasis on chance, the Atomists also stressed the factor of necessity, the universality and permanence of the properties of matter and the laws of nature. They indicated the existence of definite constraints that limit the possible combinations of atoms. Bodies are formed, they said, each of its "own" atoms, and not everything can be produced from all else (Lucretius 1956:9–11). Aristotle, acknowledging the importance of the two factors of chance and "material necessity," nevertheless criticized the atomists for being unable to offer an adequate explanation of the development and behavior of living organisms. According to Aristotle, vital processes are also driven by "final causes," involving the accomplishment of specific natural ends. Indeed, the atomists did not regard inanimate and animate objects as basically different, and so did not postulate a specific principle that pertains to the living world. Their belief in spontaneous generation was founded on their materialistic worldview.

The restricted framework of our discussion allows us only a cursory examination of some of the ideas presented by the two great Greek philosophers Plato (?427–347 B.C.E.), and Aristotle (384–322 B.C.E.). What should be noted in the context of our discussion is Plato's dualistic conception, according to which matter and form, body and soul, are two different categories—a conception that clearly conflicts with the materialistic, antiteleological monism of the atomists (McMullin 1985:7). Plato did postulate the existence of elementary units, the atoms, as constituents of the material world, but he explained their properties and the changes they undergo when they combine in terms of geometrical principles. In contrast to the atomists, Plato regarded the movements of material elements in space as an insufficient explanation of the existence of functional organization of the world as a whole and of its living creatures (Cornford 1957; Toulmin and Goodfield 1982:75–82). The creation myth, as told by Plato in the dialogue *Timaeus*, describes the shaping of passive matter by the hand of a divine intelligent craftsman, the Demiourgos, using an ideal mold of an eternal world of universal ideas. The created world, resembling a huge organized creature, as well as each of the organized beings within it, is made possible only through design, through the imprinting of a principle of reason onto physical matter (Cornford 1957:33 [*Timaeus* 30a]).

ARISTOTLE'S CONCEPTION OF THE ORGANISM

Aristotle's ideas concerning the processes of organic generation and growth and spontaneous generation are of great intrinsic philosophical interest, but they are also of enormous historical significance. His views became the predominant dogma for philosophers and naturalists, reigning until the seventeenth century. Over the centuries Aristotle's concepts were interpreted, reinterpreted, and changed by various schools of philosophy, being of special interest to Christian theologians. Therefore it is essential in discussing Aristotle to distinguish between his original philosophy of nature and later versions of it.

Aristotle's explanations of natural processes stemmed from his intimate, extensive knowledge of living creatures and from his deep interest in biological phenomena. He characterized living beings first and foremost as organized wholes in which each and every part contributes to the activities and preservation of the whole (Aristotle 1952a; Aristotle 1991; Nussbaum 1978:76–80). Another fundamental Aristotelian characterization emphasizes the orderly and regular fashion in which the organic whole develops, from its embryonic stage until it reaches its mature, functioning form (Aristotle 1952b). The regularity and level of order manifested in embryonic development, as well as the complexity and functionality of the mature organization, cannot, in Aristotle's view, be explained exclusively by random movements of atoms and the material constitution of the organism. This was the claim of his rivals, the materialists (Aristotle1961, 8:36–37). In addition to the necessary materialistic explanation, Aristotle insists on the principle of form, or the "soul," which he considers the organizing principle of the living body (Aristotle 1991:412a-413a). Unlike Plato, Aristotle does not regard the soul, this form-giving principle, as existing separate from the body. Generally speaking, unlike Plato, he does not postulate the existence of a separate realm of universal ideas. Yet in similar fashion to the general Greek worldview, Aristotle regards the basic principles of organization that are characteristic of each species as unchanging and eternal (Cooper 1987). The organic form is always embodied in an individual organism, and its eternal existence is guaranteed through the processes of procreation and heredity in which the form, the soul, is transferred from parent to offspring, from generation to generation (Aristotle 1952b: book 2, chapter 1, 731b).

In the process of sexual generation and embryonic development, matter is organized not randomly but according to its specific form—the form of the species to which an organism belongs. The form defines the essence of each organism and is also the end of its growth process. The actualization or fulfilment of this form is attained when the living creature reaches maturity. Thus, it is Aristotle's conviction that in order to fully describe and understand the processes of biological change both the principle of form and the factors that determine material movement are essential (Aristotle 1952b: book 2, chapter 4).

In his biological writings, such as *On the Generation of Animals*, *On the Parts of Animals*, and *Historia Animalium*, Aristotle discusses the phenomena connected with sexual generation and also many cases of spontaneous generation, in which organisms are formed "automatically," with no parents present. Plants, worms, insects, molluscs, fishes, and eels are spontaneously generated from inorganic and organic materials (Aristotle 1952c: book 5, chapter 1; book 5, chapter 32; Aristotle 1952b: book 1, chapter 1). Because of Aristotle's enormous authority, these descriptions served for ages as the most solid evidence for the existence of spontaneous generation. It is thus important to recognize a basic difficulty involved when the idea of spontaneous generation is discussed as part of Aristotle's philosophy of nature. For him, the generation of life, as manifested in sexual generation and embryonic development, is a teleological, purposeful, process. It is the strongest and clearest example of a purposeful process, proceeding in a nonrandom way, repeating itself and leading, unless an obstacle interferes, to the materialization of its end. On the other hand, the phenomena of spontaneous generation seem to result from the arbitrary material circumstances at a certain place and a certain moment. Here, the continuity of species cannot be guaranteed, since parents of a given species are not giving birth to their own kind (Gotthelf 1987:241).

Indeed, there is an ongoing dispute among Aristotle's interpreters as to whether the concept of spontaneous generation can be incorporated easily into his general teleological framework, or whether it diverges from this framework, representing chance-like natural phenomena (Lennox 1982). While we cannot treat this complex theoretical dispute in depth, it is important to point out certain parallels Aristotle does find between spontaneous generation and sexual generation. How does Aristotle characterize the organization of the embryo into a mature organism? In *On the Generation of Animals*, he says that in every act of sexual generation the male parent contributes the soul—the principle of form—and the female parent contributes the bodily material. The soul is a system of capacities, specific to each species, that passes from generation to generation. Though the soul is not material, it is nevertheless carried, or supported, by hot air—or *pneuma* in Greek—a material component of the male semen, endowed with a specific pattern of movements. This pattern of movements gradually organizes the material contributed by the female until the process culminates in the mature organism (Aristotle 1952b: book 2, chapter 3; book 3, chapter 11).

Here we reach the interesting point of comparison between sexual and spontaneous generation. The *pneuma*, says Aristotle, is similar in its constitution to the "fifth element" (the quintessence), out of which the heavenly bodies, including the sun, are made (Aristotle 1952b: book 2, chapter 3). It is on the basis of this similarity that we can understand, according to Aristotle, how both the heat of the sun, in spontaneous generation, and the heat of animals, expressed in the *pneuma*, are able to generate animals. Since the fifth element is distinct from the ordinary four terrestrial elements, the heat of terrestrial fire, says Aristotle, cannot generate life. Like the organization of the embryo by the hot air in the semen, the heat of the sun, warming bodies of water and the moist earth, is responsible for the organization of matter in the spontaneous generation of simple organisms.

We have noted how the atomists justified their belief in spontaneous generation, basing it on their materialistic conception. In a similar fashion, Aristotle attempts to provide a theoretical explanation of the "observed facts" that is compatible with his philosophical system. He thus relies on both material causes and an organizing factor in the processes of sexual as well as spontaneous generation. Yet in distinction to embryonic development, which, unless hindered, is internally guided, in spontaneous generation, the development and direction of the process are determined also by external and contingent circumstances. This distinction creates a tension within the Aristotelian system and is the source of the varied interpretations of the philosophical status of spontaneous generation (Gotthelf 1987:223; Lennox 1982; Cooper 1987:247). Despite the many interpretations, it is clear that the frequent descriptions of spontaneous generation in Aristotle's writings attest to the widely held, uncontested belief in the possibility of the formation of life in this manner.

2

SPONTANEOUS GENERATION— UPS AND DOWNS

The influence of Aristotle's ideas on the Greco-Roman world and on medieval thought underwent many changes in the course of the centuries. His writings were lost, rediscovered, translated into many languages, and given various meanings. His views on the phenomena of life, especially his concepts of "matter" and "form" (or "soul"), were increasingly interpreted in mystical terms. For Aristotle, the soul represented the organization of a particular material body. Later thinkers, especially Christian theologians, regarded Aristotle as a dualist, who claimed the separate existence of matter and form. Indeed, for hundreds of years, until the seventeenth century, all living creatures were thought of in terms of form and matter, and each act of generation, whether sexual or spontaneous, was seen as an act of form-imprinting, executed directly or indirectly by God (Jacob 1982:20–24; Toulmin and Goodfield 1982:91–106).

IMPRINTING FORM INTO MATTER

The shifting interpretations of Aristotle's concept of the *pneuma*, due mainly to the influence of the school of Stoicism in the ancient world, serve as a good example of the fate of many of his ideas. In the third century B.C.E., the Stoics, following Aristotle, still considered the *pneuma* as a material element that organizes living creatures. During the next few centuries, however, their views changed radically. The *pneuma*, which became known by the Latin term *spiritus*, came to be seen as a World-soul that enters an organism at birth and departs at death to join the celestial reservoir of souls (Toulmin and Goodfield 1982:96–101). In the early centuries C.E., the general belief was that the creative force of the *pneuma* is responsible for spontaneous generation. A version of this view was also suggested by the Hellenistic philosopher Plotinus (203–262) and by the Neoplatonists, who claimed that living creatures are formed from the soil and decayed organic materials by a spirit that "makes them live" (*vivere facit*) (Oparin 1968:14).

A belief in spontaneous generation was also expressed in the writings of the Fathers of the Christian Church. They faced the problem of accommodating claims of ongoing spontaneous generation to the biblical account of the divine creation of the various kinds of living beings. For hundreds of years, the predominant Christian interpretation was that of Saint Augustine (354–430), who is considered by most theologians, Catholics and Protestants alike, to be the father of Christian theology. According to Augustine, continuing spontaneous generation is made possible by a divine decree, issued at the moment of creation and active forevermore, under whose power the generation of living creatures from the earth and various organic materials persists. Augustine referred to biblical commands such as "Let the land produce vegetation," "Let the water teem with living creatures," and "Let the land produce living creatures according to their kinds." At the first event of creation, he said, God formed the "seed-principles" (*rationes seminales*) of all the species, which are the potentials of all living creatures to come. These potentials materialize throughout the history of the Earth according to a preordained plan, in both sexual and spontaneous generation, when the circumstances are favorable (Augustine 1982 [415]; Portalie 1960; McMullin 1985:11–16).

The Augustinian doctrine of spontaneous generation prevailed until the beginning of the thirteenth century, when a new dogma based in part on Aristotle's ideas gained dominance. After the fall of the Roman Empire in the fifth century C.E., knowledge of the Greek language declined in the West, and much of Aristotle's work, except for the few books that were translated into Latin, was unknown. In the Byzantine Empire and in the great centers of Islam, on the other hand, Aristotle's works were still read in the original and in Arabic, and Aristotelian philosophy was highly influential.

In the twelfth century, Aristotelian texts in Arabic and Hebrew translations were returned to the West, as were versions in the original Greek, which reached England and France from Constantinople. All these works were translated into Latin and gradually became the main source for authoritative ideas about nature. It was the Italian theologian and philosopher Thomas Aquinas (1225–1274) who grounded Christian theology on the ideas of Aristotle (Toulmin and Goodfield 1982:139–140).

Formulating his position on the creation of the various kinds of living beings, Aquinas wished to combine the seed-principles of St. Augustine with Aristotle's philosophy. Augustine upheld both the original creation of the principles and their gradual realization in time. However, since Aristotle saw the forms of all species as eternal, there was no place in his system for the creation and gradual development of living kinds. The Christian solution, suggested by Aquinas, was to regard each species as created at a certain moment, and as unchanged, or "eternal," from that moment onward. Though embracing the teachings of Augustine in part, Aquinas tended to emphasize God's direct intervention, especially in the creation of the highly developed animals (Thomas Aquinas 1945 [1265]; McMullin 1985:16–18). By the fourteenth century, Christian theologians had completely rejected any residue of an Augustinian interpretation. They regarded nature as totally subject to God's direct intervention, and interpreted Genesis as a series of divine miracles. The gradualistic Augustinian conception was now replaced by the assumption that God can foster new beginnings of each and every species (McMullin 1985:20).

It should be made clear, however, that despite these controversies, during the eras of Augustine and Aquinas and for centuries thereafter, no doubt was raised as to the possibility of spontaneous generation. This harmony between the belief in spontaneous generation and various Christian doctrines is noteworthy, especially since it tends to be forgotten following the radical change in the church's attitude toward this issue that took place at the end of the seventeenth century. Until that change there was no reason—neither in the way organisms were thought of, nor in the predominant philosophy of nature—that would prevent the church from supporting the widespread belief in spontaneous generation. Various factors that became influential during the second half of the seventeenth century, turning this belief into a blasphemy, will be discussed later.

FANTASTIC STORIES

The belief in spontaneous generation united people of all ranks of life, naturalists and lay people alike, and its intensity is evident in many fantastic stories. Alexander Neckham, an English naturalist of the twelfth century, gives a detailed report of geese being generated from the resin of pine trees under the influence of sea salt. This widely entertained notion enabled devout Christians to eat these birds as "fish" on fast days, when meat was forbidden. Ultimately, Pope Innocent III prohibited this practice. The "goose tree" legend was probably based on the observation of a species of barnacles that attach themselves to rocks, bottoms of boats, and trees and develop a calcareous membrane resembling a shell. Such creatures are found on the shores of northern Scotland and Ireland at the same time of year when polar geese arrive from the north (Oparin 1953:8–9). It seems that a lively imagination combined with a deep belief in spontaneous generation (and maybe also a craving for meat on Fridays) engendered the link between the shells on the trees and the polar geese.

Originating in the first century, the story of the formation of an embryo of a little man, a homunculus, by mixing male sperm with female menstrual blood was widely believed for hundreds of years. At the beginning of the sixteenth century, the famous Swiss physician and alchemist von Hohenheim, who called himself Paracelsus, gave an exact recipe for the generation of an embryo from human sperm with the addition of human blood. Paracelsus claimed that a special "life force" in the sperm is responsible for shaping the material elements out of which the living body is formed. This life force, according to Paracelsus, can be manipulated using certain rules of magic (Oparin 1953:9–10).

In the seventeenth century, a series of experiments on spontaneous generation by the Flemish physician J. B. van Helmont became widely known. Van Helmont is remembered today for his important studies on the role of water in the growth of plants, and for his discovery of several gaseous compounds—he was the first to coin the term "gas" (Toulmin and Goodfield 1982:150–156). Van Helmont claimed to have sealed a sweaty shirt together with grains of wheat in a vessel for twenty-one days, producing mice that resembled ordinary field mice! He believed that human sweat, like other substances from living sources, contained a nonmaterial, form-giving principle that was capable, with the help of the gas produced by the fermenting wheat, to generate life (Oparin 1953:10–11).

THE ORGANISM AS A MACHINE

Van Helmont and several other naturalists active at the end of the sixteenth century and during the first half of the seventeenth century—among them, the English physician and anatomist William Harvey, the first to demonstrate the circulation of blood—represented a transitional stage in the attitude toward spontaneous generation. On the one hand, they were still deeply influenced by the Aristotelian tradition, especially its later interpretations, and hence they attempted to explain the generation of life and embryonic development in terms of purposeful or divine forces that organize matter. On the other hand, they performed detailed experiments, and their discoveries helped to foster the mechanistic worldview that would characterize the approaching Modern Age (Toulmin and Goodfield, 1982:140).

A striking example of the conceptual changes that occurred during the seventeenth century and of their application to the issue of spontaneous generation is to be found in the ideas of René Descartes (1596–1650), the great French philosopher and scientist. He believed unconditionally in the spontaneous generation of plants, insects, and worms from decaying materials under the heat of the sun. Nevertheless, he rejected the explanations based on a life-creating principle proposed by the medieval Scholastics, and by Paracelsus, van Helmont, and other naturalists (Descartes 1983 [1644], vol. 3; Descartes 1965 [1637], part 5; Toulmin and Goodfield 1982:156–158). A living organism, according to Descartes, is a complex machine, and thus must be explained by mechanical principles operating on particles of matter. The processes of sexual generation, embryonic development, and spontaneous generation could be explained, he thought, as the working of heat and motion on each part of the "machine" as it develops in its turn (Jacob 1982:53–54).

Within the same mechanistic framework, the English philosopher Francis Bacon supported the possibility of spontaneous generation, regarding this process as evidence of the continuous connection between the inorganic and organic realms (Bacon 1879). It should be noted that, unlike previous thinkers before who conceived each life-generating act as a separate event requiring divine intervention, both Descartes and Bacon saw the generation of every organism, sexually or spontaneously, as an expression of the universal laws of nature. This attitude was part of the general worldview established by the Scientific Revolution. The tendency now was to unite the whole of nature, inanimate and animate alike, through the universal laws of mechanics (McMullin 1985:21–27). In explaining phenomena in both the physical and living realms, the role previously given to teleological causes and divine intervention was rejected. For the mechanists of the seventeenth century, it was Aristotle who symbolized the teleological shackles from which the new physics had to be set free in order to become a true science.

According to the all-embracing mechanistic conception, there was no difference, in principle, between living beings and other physical objects except in their arrangement of matter. Thus, for Descartes and other thinkers, spontaneous generation did not require any involvement of a nonmaterial factor. From a philosophical point of view, this approach takes us back to the position of the Greek atomists, who considered the motions and interactions of particles of matter as solely responsible for the generation of all natural entities. Comparison with Greek materialism is justified despite the fact that, unlike the Greek conception of the world as eternal, belief in the creation of the world and its law was inherent to the thinking of the seventeenth and eighteenth centuries. However, the new mechanists, by adopting a deistic conception, pushed the act of creation back to the origin of things and greatly limited God's intervention since then. According to the mechanistic conception established by the Scientific Revolution, the material universe can be compared to a huge clock originally made and set in motion by God, and left to function uninterrupted according to fixed laws (Toulmin and Goodfield 1982:167–168).

The realm of the new physics comprehended the whole of nature. For Descartes, all organisms, all the various animals and the human body, were machines of differing complexity. Only the human soul—the mind and free will—was set apart from the material category and the reign of physics (Descartes 1958 [1641], chapter 6; Descartes 1965 [1637]; Brooke 1991:128). Unable to account for the connection between the two categories of mind and body, Descartes left to future generations the legacy of dualism. His extreme mechanistic philosophy brought about the reaction of the animists (anima—soul, in Latin) at the end of the seventeenth century. The animists wished to explain all biological phenomena, not only the human mind, by way of the life-giving principle of the soul. Scientifically, this movement represented a retreat to Aristotelian concepts. Moreover, unlike Aristotle, the animists of the eighteenth century endowed the concept of soul with religious connotations. Yet at the same time, their efforts reflected an increasing awareness of the complexity of biological organization and of the difficulty of accounting for vital phenomena that display direction and purpose (especially embryonic development) solely on the basis of mechanical interactions (Jacob 1982:36–40).

The inability of Cartesian philosophy to deal with major organic phenomena is reflected in the position taken by William Harvey, a prominent representative of the mechanistic worldview. It is conceivable that Harvey would not have discovered the workings of the circulation of the blood had he not adopted the mechanistic conception. He described the bodily network of arteries as a hydraulic system, and the heart as a pump. And yet Harvey was not ready to give up the Aristotelian view of the organism as an organized, purposeful whole. He did not find mechanistic principles and concepts sufficient to account for embryonic development or spontaneous generation. Every act of generation, thought Harvey, must start with an "egg," matter already organized to some degree. In sexual generation it involves the fertilized egg, in spontaneous generation, putrefying meat, rotten plants, excrement, or the pupa and chrysalis of insects. From this basic organization, a purposeful process must follow. Lacking theoretical concepts to replace or complement the mechanistic ones, Harvey still spoke about "the finger of God" as the driving force behind organic development (Harvey 1847; Toulmin and Goodfield 1982:144–150; Jacob 1982: 34–35, 52–53).

THEORETICAL ALTERNATIVES—"PREFORMATION" AND THE "PREEXISTENCE OF THE GERMS"

During the first half of the seventeenth century, Descartes and other mechanistic philosophers maintained that the push and pull of material particles were enough to organize passive matter and generate life. Toward the end of the century, however, the use of the microscope and the accumulation of anatomical findings cast serious doubts on this notion. Matter was still regarded as passive, and the mechanical laws of nature were still considered to be universal and permanent, but the organism was now conceived in a different light, and consequently, so were the processes of sexual generation, embryonic development, and spontaneous generation. Observations through the microscope, especially of insects, revealed elaborate inner structures, and the metamorphosis of insects was becoming known. By dissection and examination, the "testicles" of viviparous females were found to contain what were considered to be eggs. (In the nineteenth century it became clear that they were actually follicles surrounding the true eggs.). The male semen was found to contain "animalcules," little creatures swarming everywhere, similar to the creatures observed in ditchwater and other liquids.

Descartes had explained sexual generation as the simple mechanical aggregation of material particles from male and female parents. The same mechanism was thought to explain spontaneous generation. However, when it became obvious that the organic form is extremely complex, this idea seemed inconceivable (Jacob 1982:55–57). At the same time, all previous solutions, including animistic attempts to explain sexual and spontaneous generation on the basis of mysterious forces, were also unacceptable. During the second half of the seventeenth century a new notion emerged to account for the development of organic form, a notion that was compatible with both the mechanistic and religious views and that rejected entirely the possibility of spontaneous generation. This new proposal was based on the combination of two theories: "preformation" and "the preexistence of the germs" (Farley 1977:11–12).

According to the theory of preformation, the process of embryonic development is simply the mechanical unfolding of the germ, which is a miniature version of the mature organism preformed either in the father's sperm or in the mother's egg. This extreme mechanistic approach dealt with the difficult challenge of the development of complex biological organization by abolishing the problem. The preformists claimed that there is no need to explain how the complex form of the mature organism develops, since this form already exists, in miniature, at the beginning of the process. With the adoption of the preformist solution, however, the question obviously arose, What was the origin of the preformed miniature organism? Where did it come from? (Jacob 1982:59–60).

The answer to this question was provided by the theory of the preexistence of the germs, which became popular toward the end of the seventeenth century. It postulated the divine formation, at the moment of original creation, of the germs that contained the future embryos of all living creatures. Thus, the germs carrying the miniature creatures were not generated anew by their parents, but were created originally at the beginning of time, stored one inside the other, like a series of boxes or Russian dolls within the original embryo. For instance, the ovaries of Eve contained a human embryo, inside of which there was another human embryo, inside of which there was yet another embryo, and so on (Bowler 1984:53). Upon each act of sexual generation, the miniature form starts to grow and "unroll." (Interestingly, the Latin term describing this mechanical unrolling was *évolvere*. The term "evolution" thus referred to embryonic development. In *The Origin of Species*, Darwin used the term "descent with modifications" to describe the development of species; the use of "evolution" in that sense originated later in the nineteenth century, in the writings of the English philosopher Herbert Spencer.).

The supporters of this "encasement" theory can be better understood on the basis of several prevailing conceptions. The age of the Earth was thought to be no more than few thousand years. Matter was conceived as infinitely divisible, and since the theory that postulated the existence of cells of a certain size was not yet formulated, no limit was posited as to the minimal size of the "boxed" or "encased" germs (Gould 1977:204–205; Jacob 1982:61). For a long time disputes raged between "ovists," who claimed that the germs are passed from generation to generation through the female eggs, and "spermists," who wished to grant this honor to the male sperm. Several findings, especially William Harvey's studies of embryos, finally settled the case for the ovist doctrine, and thus the theory of the preexistence of the eggs became dominant (Farley 1977:12–13; Jacob 1982:53, 63–64).

The combination of preformation and preexistence theories also resulted in the complete rejection of spontaneous generation (Farley 1977:14). For the first time a theoretical alternative accounting for the origin of organic forms suggested itself as an explanation having no room in it for the spontaneous generation of organisms. This theoretical construction reflected the mechanistic dimension of the predominant worldview, and at the same time emphasized the idea of the divine creation of all existing creatures. Descartes, having no knowledge of the degree of complexity of organisms, thought it possible to explain acts of generation on the basis of universal mechanical laws. With the accumulation of facts pointing to the complex nature of living organisms, it became impossible to envisage a world governed by mechanical laws in which sudden events of spontaneous generation were taking place.

Many historical examinations attribute the downfall of the belief in spontaneous generation to the observations and experiments performed during the second half of the seventeenth century. Attention is given to the work of Jan Swammerdam and Marcello Malpighi, and in particular to the experiments of the Italian physician Francesco Redi (1626–1698). Swammerdam and Malpighi described the sexual organs and copulation of insects, and Redi performed a series of experiments to find out whether fly maggots were indeed generated spontaneously on meat or fish. Covering the meat with a fine cloth, Redi prevented the development of maggots on the meat itself, while allowing the flies to lay their eggs on the cloth. He concluded that insects cannot generate spontaneously from putrefying meat or from fish, plants, and fruits, which can serve only as a nest for the development of the insects after the eggs are laid (Farley 1977:14–15; Jacob 1982:54).

There is no doubt that such experiments had a decisive impact on the conflict surrounding spontaneous generation. Yet the very fact that these experiments were performed, as well as the willingness of naturalists to consider their results, needs to be considered within the context of preformation and preexistence theories, which suggested a theoretical and philosophical alternative. The attitude of the Christian Church to the issue of generation should also be examined within a framework that combines philosophical, theoretical, and empirical factors. As long as events of spontaneous generation were attributed to the direct or indirect action of God, the church could obviously pledge its support. Problems began to arise when Descartes and other seventeenth-century mechanists no longer had use for God to explain events of generation, relying instead on the motion of matter under suitable environmental conditions. More and more as the complexity of the organic form was realized, the dilemma for the church intensified. Ultimately an adequate solution was found in the ideas of preformation and preexistence of the germs. Only then did it become possible to be both a good mechanist and a good Christian (Farley 1977:29; Jacob 1982:62).

It should therefore be pointed out that, in distinction to the claims usually made in historical examinations of the topic, the opposition to spontaneous generation expressed by many naturalists at the end of the seventeenth century and the beginning of the eighteenth stemmed not only from the experimental results mentioned above (Farley 1977:114). Francesco Redi is known for his unwavering support of the preexistence theory, for his belief in the divine origin of the different kinds of plants and animals and the subsequent generation of organisms only from their parents. For this reason, despite the results of his experiments, he did not deny the possibility of spontaneous generation, in some cases, from live plants and living tissues, claiming that these substances contained the principle of life, originating from divine creation (Farley 1977:15). From our point of view, Redi's experimental refutation of the spontaneous generation of flies from meat can be seen as a victory of science. This is indeed the interpretation usually given to these experiments. Examined in its own context, however, the denial of spontaneous generation in this period was first and foremost an expression of the support of the preexistence theory, which combined belief in the original creation of life with the conception of universal laws governing nature. This is why, when sufficient evidence against the preformation and preexistence theories accumulated during the eighteenth century, support for spontaneous generation was again on the rise. This time, however, it did not concern the generation of complex organisms like insects, but rather those tiny creatures, the animalcules, discovered through the microscope by the Dutch investigator Leeuwenhoek toward the end of the seventeenth century.

3

THE REVIVAL OF
THE BELIEF IN
SPONTANEOUS
GENERATION

From the second half of the seventeenth century to the middle of
the eighteenth, embryonic development was explained by the theories of preformation and preexistence of the germs. As already
noted, support of these theories, as well as the results of Redi's,
Malpighi's, Swammerdam's, and others' experiments, led to the rejection of spontaneous generation. Gradually, however, empirical
data were accumulated that cast doubt on the ideas of preformation and preexistence. These data, together with new emphases in
the philosophy of nature, reversed the earlier thinking. By the second half of the eighteenth century, the belief in spontaneous generation was again on the rise.

A NEW DYNAMIC APPROACH: THE THEORY OF EPIGENESIS

Strong evidence against preformation was provided by the phenomena of regeneration, or as it was called at the time, "reproduction,"
manifested by various organisms (Jacob 1982:72). Experiments on
aquatic worms, hydras, snails, and crayfish have shown that these
creatures can regenerate an amputated part, or reproduce a whole
body from its parts. Each part cut from a polyp is capable of growing a new polyp. In a similar manner, limbs of a crayfish or a sala-
mander, or even the head of a snail, once cut off can grow again.
These newly discovered facts were incompatible with the idea of

preformation. The processes of regeneration that followed the damage caused to the organism pointed to the possibility of a true development of the organic form, and seemed to refute the claim that a mature organism can form only as the result of a mechanical unfolding of a miniature preformed within the mother's egg. In order to reconcile the idea of preformation with the facts of regeneration, increasingly bizarre claims were made, such as the idea that miniature forms of organs exist in the body ready to face any eventuality (Jacob 1982:67–68).

Further evidence refuting the ideas of preformation and preexistence came from new observations of the phenomena of heredity. According to the theory of the preexistence of the eggs, the miniature organism is passed from generation to generation through the mother, who is the main contributor to a newborn's traits. Resemblance to both parents had been attributed to nutrition and other environmental factors. Systematic observations performed during the eighteenth century showed, however, that the father's contribution is not random and that both parents contribute in a regular fashion to their offspring. Decisive evidence was produced by hybridization experiments in plants: the properties of the hybrid, even if it was sterile, were clearly a mixture of the parents' features (Jacob 1982:68–69; Olby 1966:154). The great eighteenth-century French naturalist Georges-Louis de Buffon, rejecting the idea that matter was infinitely divisible, performed a series of calculations aimed at testing the validity of preexistence by defining the size of the encased germs. Assuming that a single germ is more than a thousand million times smaller than a man, Buffon calculated that the germ of the sixth generation following the event of creation would already be smaller than the smallest possible atom (Jacob 1982:67; Farley 1977:23). This result seriously weakened the support given to preexistence.

The renewed belief in spontaneous generation was also associated with changes in the mechanistic philosophy of nature at the beginning of the eighteenth century, especially the growing influence of Newton's ideas. Whereas Cartesian physics, rejecting the existence of a void, viewed every physical change as a result of the rearrangement of particles of matter, the Newtonian conception, elaborating on the mechanism of Descartes, focused on the concept of the force of attraction, active in empty space between particles of matter. Though matter was still conceived as passive, the action of gravitation on bodies endowed physical nature with a dynamic dimension (Bowler 1984:64; Jacob 1982:40). This dynamic approach was applied by several of the leading naturalists of the time to the investigation of the phenomena of life, resulting in a new evaluation of the processes of sexual generation, embryonic development, and spontaneous generation.

In the mid–eighteenth century, the development of organic form was conceived in a new light, presenting a theoretical alternative to preformation. The new theory of epigenesis (*epi*, after, from Greek; referring to the sequential addition of parts) described embryonic development as the structural elaboration of an unstructured egg, and thus provided a conceptual framework capable of dealing with the accumulated data. Unlike preformation, in an epigenetic process the embryo develops from a relatively homogeneous starting point, gradually differentiating into the various parts and organs of the mature organism. In the seventeenth century, William Harvey had portrayed embryonic development in a similar manner, but his ideas were not accepted since they lacked theoretical support. (For the sake of historical accuracy, it should be pointed out that Aristotle's theory of embryonic development was also epigenetic.) While epigenesis was not empirically proved until the nineteenth century, in the middle of the eighteenth century several of the most important naturalists, among them Buffon and Pierre-Louis de Maupertuis, were devoted supporters of the theory (Buffon 1749; Maupertuis 1966 [1745]).

Maupertuis and Buffon were the leading Newtonians in France. They believed that like inanimate objects, which are constructed of atoms, living creatures are also composed of invisible organic units. Forces of attraction among those units are responsible for the organization processes in the living body. Newton postulated the existence of gravitation, a mysterious force the essence of which is incomprehensible, a force that acts at a distance on bodies in empty space. Though its essence is unknown, the effects of gravitation can be formulated in a law, being regular and amenable to analysis and measurement. By analogy to the Newtonian force, Maupertuis and Buffon raised the possibility of other natural forces active in living organisms (Jacob 1982:76–77). Every paternal and maternal germ, they said, contains a variety of organic particles, which, guided by the attraction among them, create the organs. It is thus not a miniature organism that is being transmitted in heredity, but particles that produce the mature organism epigenetically (Buffon 1749:24–32; Maupertuis 1966 [1745]). In order to explain the reproduction of the parents' image in the offspring, Maupertuis granted the transmitted particles a kind of "memory" by which they retain the parents' form. Buffon, likening the process to Newtonian gravitation, spoke of a *moule intérieur*, an inner mold, the essence of which is unknown but which nevertheless organizes the particles in the growing embryo in a specific manner (Farley 1977:23).

Maupertuis and Buffon assumed the existence of organic particles characteristic of living matter and of natural forces responsible for vital processes. Their views can thus be described as "vitalistic." Vitalism is a philosophical position that distinguishes living beings from other physical objects by their unique material constitution and active forces. Buffon's and Maupertuis's Newtonian vitalism was of a special kind (Mclaughlin 1990:20–24). It was materialistic and mechanistic, lacked religious connotations, and aimed to explain the unique properties of life by natural laws, establishing a scientific inquiry of vital phenomena. The connection between this version of vitalism and the establishment of biology as a scientific discipline toward the end of the eighteenth and the beginning of the nineteenth century is worth noting. Newtonian vitalism focused on the unique characteristics of life, in particular, those of biological organization, and at the same time it drew on the concept of natural force whose effects can be investigated and measured. Through the work of naturalists and embryologists, it led to the rise of the science of biology. This vitalism, claimed François Jacob, "was as necessary for the establishment of biology as mechanism has been for the Classical period" (Jacob 1982:92).

The belief in spontaneous generation was revived with the rise of epigenetic ideas. Assuming that processes of organization were at work in the embryo, there was no reason to deny that similar activity within inanimate matter could generate life spontaneously. In distinction to the past, however, this mode of generation was now assumed to apply not to complex creatures like insects, but only to microscopic ones. Buffon devoted lengthy discussions to this possibility and described a regular cycle in nature in which organic particles released upon the death and decomposition of various organisms combine through attraction forces acting among them, forming new organisms, especially microorganisms (Farley 1977:23–24). When such a process takes place under certain circumstances within a living body, it could answer a long-debated question that puzzled even the opponents of spontaneous generation: What is the origin of parasitic worms that are found in the gut and other internal parts of higher organisms? Life cycles involving drastic changes in parasites' bodies, as well as the passage of parasites from one host to another, became known only in the nineteenth century. Until then, the available explanation for their appearance was spontaneous generation out of the host's tissues (Farley 1977:18–19).

Buffon's ideas about spontaneous generation included many references to the history of the Earth and to the development of a variety of life forms. In several of his monographs he discussed the · epochs of the Earth, during which conditions made it possible for organic particles to be generated spontaneously and to assemble spontaneously to form the first specimens, the "founding fathers" of the basic types of animals and plants. Each type, or "family" in Buffon's terminology, was characterized by a specific inner mold that passed from generation to generation, preserving the stability of the type. The changes in the type that took place with time were thought to have been caused by changes in environmental conditions, culminating in a deterioration of the original type and its divergence to different species (Bowler 1984:71–72; Buffon 1962 [1778]).

In a similar manner, Maupertuis described the spontaneous generation of the biological species on Earth in different historical periods as resulting from the combination of organic particles. Unlike Buffon, who limited the number of possible spontaneous combinations to those compatible with the inner molds, Maupertuis emphasized the random nature of material combinations. Only the successful combinations survived, he claimed, while the unfit became extinct (Bowler 1984:67). Yet it should be remembered that Maupertuis also assumed a sort of "memory," guiding the organic particles into a complex, functioning biological organization that reconstructs the species form.

EMPIRICAL AND PHILOSOPHICAL CONFLICTS

In the eighteenth century, the confrontation between ideas of preformation and preexistence and those of epigenesis provided the theoretical background for discussions of spontaneous generation. At the same time, conflicting empirical evidence was furnished by two leading naturalists, an Italian, Abbé Lazzaro Spallanzani, who wished to prove the impossibility of spontaneous generation even of tiny creatures in solutions, and an Englishman, Abbé John

Turberville Needham, whose goal was just the opposite. Interestingly, each claimed to have proved his own hypothesis despite the fact that both performed similar experiments.

Spallanzani and Needham heated organic solutions, such as meat gravy, in sealed vessels, and checked them for growth of microorganisms. Spallanzani, in addition to meticulously boiling the gravy, heated the air contained in the sealed vessels. Under such conditions, as distinct from experiments in which the sealed air was not heated, no microorganisms formed spontaneously in the solution (Spallanzani 1803). Needham, on the other hand, claimed that after he removed any trace of air from the sealed dishes and heated the meat infusions, microscopic creatures did appear in the medium (Needham 1748). From our present point of view, it is obvious that Spallanzani was more fastidious than Needham in sterilizing his dishes, thereby producing different results. However, the eighteenth-century conceptual framework within which both of them worked was not yet ready for an empirical refutation of spontaneous generation. Needham claimed that the intensive heat used by Spallanzani destroyed the "vegetative force" residing in organic material and active in the spontaneous generation of organisms. Spallanzani performed various experiments with the object of refuting the existence of the vegetative force, but it was clear that no experiment could either prove or disprove such a force (Jacob 1982:82).

The controversy between the two naturalists reflected two different approaches to nature: traditional mechanism, represented by Spallanzani, adhering to ideas of preformation and preexistence in order to circumvent the need to account for the development of the organic form, versus its rival, the new mechanism, represented by Needham, according to which organic forms develop through the dynamic processes of epigenesis and spontaneous generation (Farley 1977:26–27).

During the last decades of the eighteenth century, the attitude toward spontaneous generation in different European countries reflected local interests and circumstances. In France, support for spontaneous generation was on the rise following the growing influence of the materialistic ideas of the French *philosophes*. Denis Diderot, Julien de la Mettrie, and the Baron d'Holbach, who advanced the materialistic ideas of the Enlightenment in France, were firm atheists. They rejected the deistic combination of a mechanistic approach and the postulate of divine creation of the physical universe and all living forms. Instead they envisaged processes of cosmic development, which included repeated events of spontaneous generation resulting from random combinations of material particles. They believed these processes to be true even for the more complex animals. According to the *philosophes*, Needham's experiments as well as the phenomena of regeneration were evidence of the unlimited creativity of material nature (Diderot 1966 [1769]; D'Holbach 1970 [1770]). These views established in France a clear connection between the support of spontaneous generation and atheism, materialism, and political radicalism. This connection proved crucial to the unfolding of the Pasteur-Pouchet conflict in the nineteenth century (Farley 1977:28–29).

Rejection of preformation and preexistence and rising support for spontaneous generation took place not only in France but also in Germany during the last years of the eighteenth century and the early decades of the nineteenth. Yet in Germany these changes stemmed from an idealist, romantic philosophy of nature, the movement of *Naturphilosophie*, which was opposed in spirit to French materialism. The simplistic mechanism advocated by Diderot and other French intellectuals described the continuity of nature exclusively in terms of matter in motion, disregarding distinctions between the inanimate and the animate. The German romantic movement, on the other hand—led and inspired by the great poet, writer, and naturalist, Johann Wolfgang von Goethe, the idealist philosopher Friedrich Schelling, and the naturalist Lorenz Oken—described the unity of nature, both biological and physical, in organic terms (Lenoir 1982; Oken 1847). The world was conceived as a growing organic entity, manifesting the activity of a general spiritual principle (*Geist*), for which the epigenetic, purposeful development of the embryo served as a model (Bowler 1984:99–101). Within this worldview, events of spontaneous generation were an integral part of the general scheme of things (Farley 1977:31–34).

Throughout the eighteenth and nineteenth centuries most English scientists, unlike their French and German counterparts, rejected the possibility of spontaneous generation, mostly on theological grounds. The intellectual movement of natural theology, which claimed to have proved the existence of God based on the order and adaptation manifested in nature, was especially strong in England (Paley 1970 [1802]). With the growing awareness of the highly complex nature of biological organization, including that of microscopic creatures, English naturalists felt even more strongly that such an organization could not conceivably have been produced by mechanical processes in inanimate matter (Farley 1977:43–45). An important exception was Erasmus Darwin, Charles Darwin's grandfather, who in his poem *The Temple of Nature*, published in 1804, described the spontaneous generation of "the first specks" of life beneath the waves of the primordial sea (Darwin 1804).

The main drama during the nineteenth century concerning spontaneous generation was played in France, involving scientific, religious, and political factors. We have drawn attention in our previous discussions to the church's full support of spontaneous generation during long historical periods, and to its attempts to reconcile biblical tenets with the widespread belief in this doctrine. This state of affairs changed in the last decades of the seventeenth century, when, for empirical and philosophical reasons, preformation and preexistence became more congenial to church dogmas. Denying spontaneous generation on the basis of these theories, the church could focus on the original divine creation of all living creatures (Farley 1977:29). At the same time, it was still possible to adhere to the mechanistic philosophy of nature. The passage of "packaged" embryos from generation to generation and their mechanical unfolding upon birth were interpreted as an expression of universal mechanical laws. Toward the end of the eighteenth century the opposition of the church to spontaneous generation grew even more fierce. Following the French Revolution with its political changes, this opposition became a touchstone of loyalty to the political establishment and to the Christian faith.

In addition to its association with atheism and political radicalism, spontaneous generation was connected in France with the ideas of "transformism," the term used for biological evolution. This connection was especially fostered by the heated controversy between two renowned French naturalists, Georges Cuvier (1769–1832), one of the giants of science at the time, who vehemently opposed any mention of evolution, and Jean-Baptiste Lamarck (1744–1829), who, fifty years before Darwin, formulated his own theory of evolution (Mayr 1982:363–367, 343–351). Cuvier, the founder of paleontology and comparative anatomy, emphasized in his work the harmony and purposiveness of biological organization. He was fiercely opposed to the idea that forms of life could have developed into different ones, and accepted as self-evident the original creation of all basic organic forms. Not ready to view the organized biological whole as a product of material rearrangements, Cuvier, claiming that life always generated out of previous life, supported the theories of preformation and preexistence (Cuvier 1863 [1817]; Farley 1977:39). He regarded Lamarck, who advocated transformism, as his archenemy.

Lamarck's theory shared with Darwin's the basic tenet concerning the evolution of biological species during the history of the Earth (Lamarck 1963 [1809]). In several of his main claims, however, Lamarck differed from Darwin (Mayr 1982:358–359). Unlike Darwin, he assumed repeated events of spontaneous generation of the simplest life-forms through a materialistic, nonmysterious process. (Bowler 1984:79). He spoke about the existence in nature of "subtle fluids," which, by carving channels through a homogenous, gelatinous inorganic matter, could, under certain conditions, organize this matter to bring about the first organisms (Mayr 1982: 357). These simplest forms, continuously generated in such a manner, are the basis of several types of evolutionary scales. Along these scales the more complex types of organisms gradually develop, each organism being pushed forward by an inner drive toward higher complexity and active adaptation to changing environmental conditions. Both the inner drive toward complexity and its expression in better adaptation to the environment were again accounted for by Lamarck by the physiological activity of different fluids in the animate matter (Mayr 1982:354). It was through Lamarck's writings that the connection between transformism and spontaneous generation was established in France.

During the second and third decades of the nineteenth century, public debates between Cuvier and a fierce opponent of his theories, the biologist Étienne Geoffroy Saint-Hilaire, ended with Cuvier's incontestable victory. This state of affairs—an atmosphere of political-religious conservatism during the reign of Napoleon, coupled with the inferior status of Lamarck vis-à-vis Cuvier— brought about a sharp decline in the popularity of the idea of spontaneous generation.

4

LOUIS PASTEUR— THE DEATHBLOW TO SPONTANEOUS GENERATION

In 1859 the French Academy of Sciences established a prize of 2,500 francs to be awarded to the scientist who could throw new light on the question of spontaneous generation. The prize was awarded in 1862 to the great scientist Louis Pasteur (1822–1895) on the basis of his study "On the Organized Corpuscles That Exist in the Atmosphere," published in 1861 in the *Annals des sciences naturelles* (Pasteur 1922–1939, 2:224). In this work new experiments were presented that decisively discredited the doctrine of the spontaneous generation of microorganisms. Pasteur's experiments and ideas were a most significant milestone in the history of our subject, and they feature prominently in every historical account of spontaneous generation. Pasteur is usually portrayed as a hero, who represents the victory of science over the forces of reaction, which as late as the nineteenth century stubbornly clung to outdated ideas (Vallery-Radot 1926; Dubos 1950; Oparin 1953:21). In the following discussion, this picture will be examined and called into question.

The spontaneous-generation conflict in France in the second half of the nineteenth century has served historians of science as an interesting case study of the nature of science in general. In the way that their analyses deal with the role of extrascientific factors in shaping the conceptual content of science, historians reveal that they are engaged in a conflict of their own. On one side are those who regard science—in our case, the work of Pasteur on spontaneous generation—as an objective enterprise capable of determining factual truths about the world, and who overlook cultural and political influences. On the other are those who doubt the objective, unbiased nature of science and who consider extrascientific influences to be strong enough to shape factual truths.

Though I cannot dwell on this important controversy, there is no way I can ignore it. In my view, both extreme "objectivist" and "relativist" positions have to be rejected. I will support the view that political and religious factors could and did influence the motivation for conducting experiments on spontaneous generation, the decision to publish or suppress experimental results, and especially, the interpretation of and significance ascribed to these results. At the same time, no historical analysis of the debate on spontaneous generation should ignore the factual truth that spontaneous generation of microorganisms does not occur. It should not ignore the obvious truth that no political or religious view could change the facts concerning the spontaneous generation of living creatures and that Pasteur's work, despite his extrascientific biases, reflected these facts.

A historical study published in 1974 by John Farley and Gerald Geison, followed by additional works on Pasteur's prolific scientific achievements, including his investigation of spontaneous generation, presented a radical relativist revision of the conventional heroic stories. This extensive research (see especially Farley and Geison 1974; Farley 1977:92–120) culminated in 1995 in the publication by Geison of a provocative work, *The Private Science of Louis Pasteur*, which examines the many aspects of Pasteur's vast scientific career. In their 1974 discussion of the controversy between Pasteur and Félix Pouchet over the question of spontaneous generation, Farley and Geison, to a large extent, ignored the nontrivial fact that spontaneous generation of microorganisms does not occur. They also presented an overly sympathetic picture of Pouchet, not emphasizing enough Pasteur's great scientific expertise. In his 1995 book, Geison offers a critical evaluation of Pasteur's most prominent scientific achievements. At the same time, in response to many critics of his and Farley's position in their paper on spontaneous generation, Geison acknowledges that the interpretation they adopted "was rather too crudely 'externalist' in form and asymmetrically tilted in Pouchet's favor" (Geison 1995:321).

Rejecting the more extreme relativist position of Farley and Geison, I will still draw on the many facets of their investigation, especially in my description of the nonscientific elements—religious, political, philosophical—that had a great influence on Pasteur and his rivals. In addition, examining the specific experiments on spontaneous generation, I will also give a short sketch of the wider scientific and philosophical developments that led to the conflict.

We have already noted in the last chapter that toward the middle of the nineteenth century, following the debates between the victorious Cuvier and his opponents Lamarck and Geoffroy Saint-Hillaire, and also in tune with the anti-materialist and anti-evolutionist thinking in France at the time, the idea of spontaneous generation suffered a sharp decline in popularity. The controversy concerning this question, both among scientists and the public at large, picked up again in connection with experiments conducted by the naturalist Félix Pouchet, director of the Natural History Museum at Rouen and a corresponding member of the French Academy. In a paper published in 1858, Pouchet described the spontaneous generation of microorganisms in boiled-hay infusions (liquid extracts). He denied the accusation that the origin of these microorganisms was in the air, arguing that they appeared following the fermentation of organic material in the examined infusions (Pouchet 1858). It should be pointed out that Pouchet did not claim that adult organisms were being produced spontaneously, only their eggs. It was Pouchet's major work, *Heterogenesis, or Treatise on Spontaneous Generation*, published in 1859, that motivated the academy to seek a solution to the debate by offering the above-mentioned prize. I will now examine a few of the factors contributing to the rise of the conflict between Pasteur and Pouchet.

THE SCIENTIFIC BACKGROUND—CONFLICTING MESSAGES

Various scientific and philosophical developments in Europe around the middle of the nineteenth century—some of which affirmed and some of which denied the possibility of spontaneous generation—caused much confusion on this issue. The life cycles of parasitic worms were discovered at this time. This made it possible to account for the origin of parasites without recourse to the long-held assumption that these organisms were produced by the diseased body tissues of the host. Thus, one of the strongest traditional arguments for spontaneous generation was finally laid to rest (Farley 1977:56–66). Another highly relevant debate concerned the question of the formation of new cells in an organism. Cell theory, according to which all living tissues—in plants and animals alike—consist of cells, was formulated in the 1830s. One of the fathers of the cell theory, Theodor Schwann, later claimed that new cells are produced from a liquid stuff, an extracellular material surrounding existing cells, in a process similar to the formation of crystals from inorganic material. Supporters of spontaneous generation, among them Pouchet, saw in this "free cell formation" a vindication of the idea of spontaneous generation. The enemies of spontaneous generation, making the same connection, attacked Schwann's "exogeny hypothesis" on this very ground, claiming that new cells could be formed only endogenously, from previously existing cells. The "cell continuity" thesis that gained force in the 1850s was seen by the German biologist Rudolph Virchow, its main proponent, as decisive evidence against any possibility of spontaneous generation. Virchow, who spoke in favor of a "new vitalism," stressed the unique organization of the cell and rejected vehemently any claim that a living cell could be produced spontaneously (Virchow 1971 [1858]; Toulmin and Goodfield 1982:350–354).

At the same time, however, a few of the most prominent German physiologists had founded the "Physical Society," wishing to establish physiology and biology on a physicochemical basis and rejecting any vitalistic attempt to distinguish organisms from other natural objects (Lenoir 1982:215–228). Logically, it could have been expected that such a trend would lead to support for the idea of the generation of life from inanimate matter. This indeed happened, but only later, after the publication of Darwin's theory of evolution, when the debate focused on the emergence of the first organisms on the primordial Earth. Yet the complex nature of the history of ideas is demonstrated by the fact that, at first, the "physicalist" physiologists objected to spontaneous generation because they associated this doctrine with the mystical ideas of the *Naturphilosphie* movement, from which they wished to free themselves (Farley 1977:48).

Additional confusion concerning spontaneous generation stemmed from the debates among organic chemists on the nature of organic compounds. For hundreds of years it was believed that only organisms can produce organic substances and that these substances are endowed with special "vital forces." During the first half of the nineteenth century, chemists succeeded in synthesizing various organic compounds from simple inorganic materials, and thus seemed to sever the traditional connection between vitalism and organic chemistry. And yet a few of the best organic chemists of the period still believed that the complexity of organic compounds derives from a non-physicochemical power, and as a consequence claimed that living creatures could not have been formed exclusively through the action of natural forces. Materialistically inclined organic chemists, on the other hand, saw the synthesis of organic material outside the cell as a demonstration of a possible first step in the spontaneous generation of organisms (Jacob 1982:94–95; Farley 1977:67–69).

THE SPONTANEOUS-GENERATION CONTROVERSY—AN EXPERIMENTAL ISSUE

In our discussion of the conflicting experiments of Needham and Spallanzani in the eighteenth century, we made the point that the question to be settled by the two opponents—Is there or is there not a life-generating-force?—could not be resolved experimentally. In the mid–nineteenth century the situation changed because of several factors that turned the controversy concerning spontaneous generation into an experimental issue. Most influential were the many debates on the nature of putrefaction and fermentation, traditionally considered as processes spontaneously generating microscopic organisms (Jacob 1982:97; Fruton 1972:22–86). These debates became associated with another controversy pertaining to the characterization of infectious diseases and the nature of pathological changes in diseased tissues (Farley 1977:82–87). It was much more than an academic, theoretical controversy: Europe was devastated in the nineteenth century by a series of cholera epidemics, which caused hundreds of thousands of deaths.

Many physicians and scientists believed that diseases were caused not by living organisms but by particles disseminated in the air, food, and water following the disintegration of living matter. Such particles, it was claimed, also existed in tissues of different organisms and under certain conditions might turn contagious. Others argued that diseases are caused by a chemical agent: in specific circumstances a chemical poison is produced in the body tissues that causes disease and excites the production of more poison. According to one of the best-known physicians in England in the mid–nineteenth century, Henry Bastian, in certain pathological cases chemical processes in living tissue lead to the generation of microorganisms. Microorganisms are thus the result and not the cause of disease, but they initiate additional chemical changes in the diseased tissues, causing an aggravation of the pathological state. Bastian, a professor of pathological anatomy who was a long-standing supporter of the doctrine of spontaneous generation, believed that the prevalent chemical theories about the nature of infectious diseases provided further evidence for this doctrine (Farley 1977:87).

The debates on the causes of putrefaction and fermentation in organic solutions, such as the souring of milk and the fermentation of alcohol, were another influential factor in preparing the scene for the experimental resolution of the issue of spontaneous generation. It was not clear whether yeast, known to be involved in these processes, was a chemical agent or a living organism. Even assuming that yeast was an organism, it had not been decided whether fermentation was a chemical process, leading to yeast's spontaneous generation, or yeast was a microorganism, existing in the air, able to contaminate organic solutions and cause fermentation. Under these circumstances an alliance evolved between those who wished to prove that microorganisms, originating in the air, were the cause of fermentation and those who wished to refute the idea of spontaneous generation (Farley 1977:48–51). Through the examination of two questions—whether the air indeed contained microorganisms, and whether through contact between the air and organic liquids fermentation would occur—the validity of the doctrine of spontaneous generation could be experimentally tested.

In his paper of 1858, as we have seen, Pouchet joined this debate, claiming that microorganisms appeared in boiled-hay infusions exposed to artificially produced, hence sterile, oxygen or air. He concluded that these microorganisms were spontaneously generated. Pouchet's extracts were made from boiled hay, a procedure supposed to secure the destruction of preexisting organisms. To prevent contact with external contaminants, the infusions were isolated in a vessel made of mercury and kept under mercury. At the same time they were exposed to artificially produced, germ-free air, since it was commonly believed that oxygen is necessary for arousing life in organic extracts. In response to the criticism that he did not destroy all the microorganisms in his solutions, Pouchet expressed his disbelief that living organisms could survive boiling temperatures. He denied the charge that the air contained a large number of germs that had somehow infiltrated his solutions, saying that the particles found in the air were not germs but grains of wheat and granules of silica. Pouchet concluded that in the presence of oxygen the processes of fermentation in the hay infusions brought about the spontaneous generation of microorganisms.

PASTEUR—FROM CRYSTALS TO FERMENTATION TO SPONTANEOUS GENERATION

How did Pasteur become involved in the controversy with Pouchet? Pasteur's studies of spontaneous generation resulted from his previous research on the fermentation of alcohol and lactic acid. These studies had, in turn, been followed his work on the properties and structure of crystals of various chemical compounds. Pasteur was continuing the work of the German crystallographer Eilhard Mitscherlich and of the renowned French physicist Jean Baptiste Biot, who had found that certain crystals, such as those of quartz, are asymmetric. These crystals have hemihedral facets—a set of facets arranged in either a right-handed or left-handed formation. Asymmetric crystals exhibit a special optical activity: they rotate the plane of polarization of polarized light either clockwise or counterclockwise. Biot had found that upon dissolving these crystals in solution, this optical activity is lost. He also discovered that certain naturally occurring organic compounds rotate plane polarized light even while in solution. He thus postulated that whereas optical activity in quartz stemmed from the asymmetric structure of the crystal, in organic compounds it was associated with the asymmetric structure of the organic molecule itself (Biot 1844).

Pasteur's research in the late 1840s focused on the crystals of various salts of tartaric acid, a constituent of grapes and a by-product of the wine industry. Based on Biot's work, Pasteur knew that these crystals have hemihedral facets, all oriented in the same direction, and that a solution of these crystals shows optical activity: the plane of polarized light is always rotated clockwise. Pasteur's most striking discovery, in 1848, concerned another by-product of alcohol fermentation, racemic acid, whose chemical composition is identical to tartaric acid, but which surprisingly does not show any optical activity. Pasteur found that racemic acid contains not one but two kinds of crystals, each with hemihedral facets oriented in opposite direction, each the mirror image of the other. He succeeded in separating manually, under the microscope, these two different kinds of crystals and showed that their corresponding solutions rotated the plane of polarized light in opposite directions. One of the kinds, it turned out, was identical to tartaric acid, and the other was its optical isomer. (Isomers are substances that share the same chemical composition, but nevertheless exhibit different properties.) Pasteur's discovery of optical isomerism made it clear why racemic acid lacks optical activity: the presence in equal mixture of both isomers—the right- and left-handed ones—cancels out their corresponding optical activities (Pasteur 1922–1939, 1:77–120).

Elaborating on Biot's ideas, Pasteur concluded that the asymmetrical structure of the crystals of tartaric acid indeed stems from the asymmetrical structure of the organic molecule itself. This phenomenon is characteristic of many other organic substances, and Pasteur's hypothesis, later to be proved correct, was that the asymmetrical crystalline structures and optical isomerism derive from a specific spatial arrangement of the atoms in these molecules. Toward the end of the nineteenth century it was shown that optical activity is manifested by organic compounds containing carbon atoms to which four different chemical groups are attached. Later still it became known that due to the character of the electronic orbitals of the carbon atom, the four different chemical groups can be spatially arranged in two mirror-image configurations, which are responsible for the opposite optical activity (Fieser and Fieser 1961: 249–260).

Pasteur could not have known the exact geometrical arrangement of asymmetrical molecules or the electronic nature of the carbon atom. Yet he considered it extremely significant that chemical reactions *outside* the living cell, which involve organic substances, always produce an optically inactive mixture of the two isomers. Chemical synthesis of an organic compound always results in a mixed product, called a racemic mixture. Biological activity within the cell, on the other hand, is specific: only one of the optical isomers is synthesized or retained. Pasteur discovered that certain microorganisms assimilate and feed on only the right-handed isomer of racemic acid, never the left-handed one. It was generally found that living organisms can utilize for nutrition only one out of a mixture of two optical isomers present in their growth medium. As a consequence of his crystallographic work during these years, and following other experiments in the early 1850s, Pasteur came to believe that molecular asymmetry and optical activity were the decisive criteria for discriminating between biological activity and other chemical processes—in fact between life and inanimate matter (Pasteur 1922–1939, 1:343). As will be shown later, this distinction was of great significance throughout his scientific career.

From 1856 Pasteur focused his attention on the study of fermentation, particularly fermentation of lactic acid—a process that produces optically active organic substances. The great chemists Berzelius and von Liebig theorized that fermentation was a purely chemical process. Pasteur, on the other hand, tended to associate fermentation with the activity of microorganisms. Since for him the most important distinction between living and inanimate matter was the presence or absence of molecular asymmetry manifested in optical activity, and since fermentation produced asymmetric products, Pasteur claimed that the cause must be biological and not chemical. On the basis of a series of experiments, he became convinced that fermentation indeed depended on the activity of living organisms. Further experiments by Pasteur indicated that different microorganisms were specifically responsible for different processes of fermentation. This conclusion lent additional support to the physiological theory against the chemical theory of fermentation (Pasteur 1922–1939, 2:51–126).

THE REFUTATION OF SPONTANEOUS GENERATION

At this stage, Pasteur had to face the question as to the origin of the microorganisms causing fermentation. He postulated that they are always present in the air and cause fermentation in solutions exposed to air. By proving this, Pasteur could deny the possibility of spontaneous generation. Moreover, the claim that the air contained a specific component or a mysterious principle, supposedly necessary for arousing life in fermenting organic matter, would be disproved. Pouchet's contention that the particles in the air were nothing but grains of wheat and granules of silica would also be undermined.

In several experiments, Pasteur aspirated atmospheric air through a tube plugged by guncotton to trap any grains of dust. On these occasions no fermentation was observed in the boiled organic infusions connected to the tube. Pasteur's hypothesis was that this procedure filtered from the air microorganisms that had been absorbed by the dust particles. This was confirmed in a series of impressive experiments in which a wad of cotton, used to filter atmospheric dust, was then inserted into boiled sugared yeast-water, previously proved to be sterile, a procedure leading to contamination and fermentation in the liquid. The decisive evidence for Pasteur's claims was provided by his famous experiments with swan-necked flasks. He filled normal flasks with sugared yeast-water, urine, and beet juice and extended their necks in curves shaped in the form of the letter S. Without sealing the necks' ends, thus leaving the various organic liquids exposed to the air through a narrow opening, he boiled the contents of several of the flasks. No contamination was detected in the boiled liquids despite the exposure to air, while in the liquids not boiled the organic material fermented and decomposed very rapidly. Pasteur's explanation was that atmospheric dust and the microorganisms absorbed by it were captured in the swan necks and could not reach the solutions. Thus no contamination could occur in the boiled organic liquids. The decisive proof was provided when Pasteur snapped off the necks and dipped them into the boiled liquids, which then showed growth of microorganisms (Pasteur 1922–1939, 2:187–191).

In other experiments, Pasteur exposed previously boiled organic liquids to open air in different environments, showing that the density of germs in the air varied with location and altitude. As mentioned earlier, during the debate between Needham and Spallanzani in the eighteenth century, Spallanzani was accused of having destroyed the life-giving force, residing in organic material by intensively heating his dishes. In order to counteract any such accusation, Pasteur devised more sophisticated experiments in which he collected blood and urine directly from the veins and bladders of animals and exposed the liquids to filtered air, proving that these liquids did not ferment even without heating (Pasteur 1922–1939, 2:170).

In light of these experiments, it is clear why Pasteur accused Pouchet of sloppy experimentation and claimed that he had inadvertently introduced contamination into his hay infusions, particularly by using dust-covered mercury. Indeed, most historians who have analyzed the experiments performed by Pouchet and his collaborators have found them to be highly inadequate. However, an important point concerning these experiments should be made. Only in the 1870s, a decade after Pasteur and Pouchet conducted their investigations, did it become clear, through the work of the German botanist Ferdinand Cohn and the English physicist John Tyndall, that Pouchet's hay infusions, unlike Pasteur's yeast extracts, contained *Bacillus subtilis*. This microorganism produces heat-resistant spores, which upon exposure to oxygen become active bacteria.

Even assuming that Pouchet's flasks were contaminated by other microorganisms, this crucial fact determined the outcome of Pouchet's experiments, and should be acknowledged in any discussion of the debate. The difference in their experimental acumen notwithstanding, since they were using different media, Pasteur and Pouchet were dealing with different microorganisms, which reacted differently to heating and boiling. This fundamental fact has not been mentioned in many of the historical analyses that focus exclusively on Pouchet's negligence and incompetence as a scientist (Farley 1977:11, 134; Vandervliet 1971:43–54). As opposed to Pouchet, Pasteur has been claimed to have represented the true, objective spirit of science, free from religious, political, or philosophical influences.

THE PHILOSOPHICAL–IDEOLOGICAL CONTEXT

Was this indeed the case? Was Pasteur an objective, unbiased researcher, while Pouchet was bound by old-fashioned beliefs? As already noted earlier in this chapter, under the scientific circumstances that developed in the late 1850s, and through his ingenuity as a theoretician and experimentalist, Pasteur was able to tackle the question of spontaneous generation and to deliver a deathblow to this perennial doctrine. Yet, both for scientists and for the public at large, spontaneous generation was far from being a purely scientific issue. Its philosophical, religious, and political connotations were far-reaching, and Pasteur could not extricate himself from this context—neither in his private deliberations on the subject, nor in his public addresses devoted to it. It is highly significant that the wider implications of Pasteur's experiments continued to reverberate long after the actual debate was over. Moreover, various signs point to the role of political considerations in Pasteur's conduct during the controversy. Most noteworthy is the fact that he chose not to publish some of his earlier, more controversial experiments until much later, when the political climate in France had changed (Geison 1995:133–138).

The attitude to spontaneous generation had become a measure of one's political-religious intentions in mid-nineteenth-century France. Whoever supported spontaneous generation was considered to be siding with the materialistic, radical, anticlerical camp that threatened public order and was associated with the liberal enemies of France in Europe. The atmosphere became even more heated

after the publication in 1862 of a French translation of Darwin's
The Origin of Species by Madame Royer, a famous materialist and
atheist, who in her preface to the book fiercely attacked the Catho-
lic Church. The close association between the theory of evolution,
or transformism as it was called in France, and the notion of spontaneous generation was fostered at the beginning of the nineteenth
century through the ideas of Lamarck. This relationship was emphasized in the 1860s, when several important French scientists
attacked Darwinism on the basis of Pasteur's refutation of spontaneous generation. Clearly, the scientific community attached wide
philosophical and scientific implications to Pasteur's experiments.
It is only against the background of this intersection of science, religion, and politics that we can appreciate the participation of non-
scientists in the debates on spontaneous generation among Pasteur,
Pouchet, and other members of the French Academy of Sciences
(Farley 1977:94–96). Pouchet, the enthusiastic supporter of spontaneous generation, was conceived by most of the public and the
scientific community as threatening the position of the church and
of the French emperor Napoleon III. In the heated scientific-
ideological discussion, the fact that Pouchet was accused of things
that he totally rejected went unnoticed.

PASTEUR AND POUCHET—A MULTIFACETED PICTURE

In *Heterogenesis*, Pouchet in fact expressed explicitly antimaterialistic, vitalistic views. He supported heterogenesis, the spontaneous
generation of life from organic matter, since this matter, according to Pouchet, was endowed with "plastic forces," or "latent life."
He rejected, on the other hand, the possibility of abiogenesis, the
generation of life from inorganic matter, because this matter, he
thought, lacked any such life-generating forces. At the beginning,
said Pouchet, there was spontaneous generation inspired by God,
and there was no reason to suppose that God had subsequently lost
his power. Furthermore, Pouchet proclaimed the repeated creations
of the different species during the history of the Earth and objected
to any idea of the transformism of species (Pouchet 1859:95–127).
Thus, in conflict with his public image in the midst of the debate
with Pasteur, and contrary to the interpretation of his views by later
historians, Pouchet maintained the most conservative positions
(Farley 1977:97–100).

And what about Pasteur? What was his philosophical position

on the generation of life from matter and his ideological attitude toward this issue ? Throughout his life Pasteur was a true believer in God and, by his own admission, a supporter of antimaterialism or "spiritualism," as it was called at the time. He was a conservative in his political views and an enthusiastic admirer of the emperor Napoleon III, who helped to promote Pasteur's research and status both morally and financially. A detailed examination of Pasteur's views on the relationship between matter and life during his long career reveals a complex and peculiar mixture of apparently conflicting elements. At several significant junctures, there is an obvious discrepancy between his private ideas on these issues and his public proclamations.

Pasteur's most notable public expression of his views is to be found in a lecture devoted to the religious-philosophical implications of the controversy over spontaneous generation. It was delivered in 1864, at the Sorbonne, to the social and intellectual elite of Paris (Pasteur 1922–1939, 2:328–356). In a highly dramatic manner, Pasteur described some of his decisive experiments, while persuasively pointing out Pouchet's experimental inadequacies (Latour 1988). Yet Pasteur introduced his listeners to the subject by stressing the deep relevance of the controversy to the great questions challenging human thought: the existence of God (in Pasteur's terms, the concept of a "useless God"), the eternity of matter, and the possibility of evolution, including the evolution of man. The question of whether spontaneous generation is possible is crucial, he insisted, because it amounts to the key question: Can matter organize itself to form a living system? (Pasteur 1922–1939, 2:328). The answer to this question, Pasteur pointed out, resolves the conflict between materialism and spiritualism. If matter can organize itself—if such a process occurred on the primordial Earth—then a divine Creator is indeed useless. In his lecture at the Sorbonne, Pasteur explicitly sided with spiritualism, claiming that life can form only from previous life—that matter cannot organize itself into a living system. Here Pasteur made it clear that taking a stand on the issue of spontaneous generation involved religious and philosophical considerations. Nevertheless, in the same lecture he claimed that he himself, as a scientist, was free of any preconceived idea on the matter, and hence completely unbiased in his research.

Even without a discussion as to whether any scientist can be completely unbiased in his research, the story that follows casts doubt on Pasteur's self-proclaimed independence of ideas external to his scientific research. Earlier in his career, while conducting his crystallographic research on molecular asymmetry and optical activity, Pasteur pondered the question of whether inorganic and organic matter can be artificially transformed into a living system. He even attempted to resolve this question in an experimental fashion. To a reader of Pasteur's standard biography, familiar with his famous experiments on spontaneous generation and with the spiritualistic, vitalistic ideas expressed in the Sorbonne lecture, these attempts may come as a surprise. It turns out that in the early 1850s Pasteur was trying to discover "a cosmic asymmetric force," a new force of nature responsible for the formation of active organic compounds and hence of life. For a few years, starting in 1852, he performed experiments in which he applied polarized light and magnets to different crystals. The object was to produce molecular asymmetry and optical activity through the use of asymmetric forces (Pasteur 1922–1939, 1:327, 341). Despite the failure of these experiments, Pasteur continued to pursue similar ideas throughout his life, even while, in the 1860s, he was engaged in proving that life cannot be generated in the test tube from organic matter, and while publicly claiming that matter cannot organize itself to form life.

COULD THE QUESTIONS OF ABIOGENESIS AND HETEROGENESIS BE SEPARATED?

Several historians of science who belong to the "objectivist" camp and focus mainly on Pasteur's experimental work, such as Nils Roll-Hansen, do not seem to see a problem here. Roll-Hansen claims that the issue at the heart of the controversy between Pasteur and Pouchet was heterogenesis—can life be spontaneously generated from organic matter here and now?—and not abiogenesis, the generation of life from inorganic matter, now or in the distant past. Roll-Hansen maintains that heterogenesis and abiogenesis were scientifically two separate questions (Roll-Hansen 1983:490–491). Pouchet, he says, supported heterogenesis and rejected abiogenesis, while Pasteur's choice was the opposite: in his controversy with Pouchet he indeed refuted heterogenesis, whereas in his experiments in the 1850s and later in life he believed in abiogenesis. The question we have to address is whether heterogenesis and abiogenesis can indeed be treated as separate scientific problems in light of the philosophical and ideological positions of Pasteur and

Pouchet. Can such a separation indeed resolve the apparent conflict between Pasteur's rejection of material self-organization and his "asymmetric forces" experiments?

The presumed separation of heterogenesis and abiogenesis evidently does not fit the case of Pouchet, who rejected abiogenesis and embraced heterogenesis on philosophical and religious grounds: being a vitalist, Pouchet, like other vitalists throughout history, assumed the existence of an unbridgeable gap between life and inanimate matter and believed abiogenesis to be impossible. Heterogenesis, on the other hand, was possible, since organic matter was endowed by God at the time of creation with "generating-plastic forces," the same generating forces that are found in living organisms. Furthermore, the association between the issues of heterogenesis and abiogenesis on philosophical grounds was the basis for the mounting opposition in France to Darwin's theory of evolution following Pasteur's experimental refutation of heterogenesis. The connection between the questions of spontaneous generation and the primordial origin of life shaped attitudes toward the Pasteur-Pouchet controversy not only in France during the nineteenth century but elsewhere for many years to come. Upon examining the case of Pasteur, it is also clear that his attitude to heterogenesis and abiogenesis was not based only on scientific considerations. Pasteur's public address at the Sorbonne in 1864, in which his experimental refutation of heterogenesis was clearly used to support spiritualism and to deny the possibility of the natural origin of life, is a strong case in point.

Roll-Hansen suggests a different interpretation of Pasteur's claims in the Sorbonne lecture. He describes Pasteur as a typical representative of "the classical liberal ideal of science," according to which scientific facts are not influenced by values, but values "have to respect the facts." In the controversy over spontaneous generation, he says, Pasteur "boldly stressed that he had approached the question without prejudices concerning atheism, materialism, spiritualism, etc." For Pasteur, Roll-Hansen contends, facts were independent of values, yet at the same time he realized that the facts he discovered concerning spontaneous generation did have strong religious implications. This is why Pasteur pointed out that the negative results of his experiments were a blow to atheists (Roll-Hansen 1988:298).

Furthermore, the contention that Pasteur the scientist approached the issue of spontaneous generation with no prejudices,

and that his positions on abiogenesis and heterogenesis can be separated on a scientific basis, seems to find support in two lectures
given in 1860. In these lectures, titled "On the Asymmetry of Naturally Occurring Organic Compounds," Pasteur expressed his vitalistic inclinations, emphasizing that molecular asymmetry and
optical activity are the demarcation line separating the living from
the inanimate. Nevertheless, on the same occasion he spoke of the
need for a physical force as the basic cause of asymmetry in organic molecules (Pasteur 1922–1939,1:314–344). On the face of it,
we witness here the possibility of the coexistence of Pasteur the
"physicalist" and the "vitalist." Yet this simple picture has to be
reexamined in light of Pasteur's published ideas on these issues
starting in the mid-1870s, and especially his writings in the 1880s.
Only in 1883 did Pasteur finally report on his provocative experiments in the 1850s. It is in these later years that he spoke of the
origin of life in terms of the application of asymmetric physical
forces to simple inorganic compounds, applying the term "spontaneous generation" in this context.

Not only are we not dealing here with two distinct scientific
questions, we are faced with a huge paradox: whereas Pasteur proclaimed that life comes always from previous life and that matter
could not have organized itself to form life through natural means,
he was trying to duplicate such a process in his own laboratory using simple inorganic molecules (Geison 1995:138).

In my view, in order to solve this paradox, we must acknowledge the influence of political and religious factors on Pasteur's public conduct during the controversy over spontaneous generation.
It is clear that in the Sorbonne lecture, Pasteur was first and foremost expressing his convictions as a true Christian believer. Con-
veniently enough these convictions suited his position as a defender
of the political system and its cherished antimaterialistic values. The
fact that he did not disclose any information about his 1850s experiments until 1883 cannot be explained by his claim at that later
time that there was nothing much to report on. As pointed out by
Geison, this fact had not changed in 1883 (Geison 1995:138). What
had changed, though, was the political situation after the fall of
the Second Empire. This change, which was much regretted by
Pasteur, nevertheless created an atmosphere that was more liberal
and tolerant of materialistic ideas.

Geison and Farley suggest a way out of the paradox by stressing the crucial distinction, in Pasteur's mind, between "ordinary

chemical procedures" and physical asymmetric forces of cosmic origin. Ordinary chemical procedures, according to Pasteur, were incapable of bridging the gap between matter and life, whereas physical asymmetric forces were the true cause of the organic "elements of life." Geison even thinks that for Pasteur God was the original source of the cosmic asymmetric force. If this is the case, then the paradox is indeed resolved, and we can see the many facets of Pasteur's conceptual world as parts of a coherent whole. Heterogenesis from organic material is impossible because it relies on chemical manipulation; abiogenetic experiments in which asymmetric forces are applied to simple chemical compounds are, in principle, possible because they are an attempt to imitate nature's true way of creating life. Pasteur's conception is also fully compatible with a religious-vitalistic attitude, since the original cause responsible for bridging the gap between matter and life—"nature's true way"—is God's creation (Farley and Geison 1974:193–197; Geison 1995:141–142).

In order to integrate the complicated story of Pasteur and spontaneous generation into our general discussion of the history of ideas about the origin of life, two essential observations are called for. First, as will become apparent in the following chapters, the true resolution of the problem of spontaneous generation became possible only in the twentieth century, when the evolutionary worldview was applied to the origin-of-life question. This resolution amounted to the simultaneous realizations that the concept of spontaneous generation was indeed dead, and that life originated on the primordial Earth through a natural, gradual process. Second, Pasteur's greatness as a scientist cannot be disputed, and it is evident that his experiments on the spontaneous generation of microorganisms finally discredited this doctrine, which had been part of humanity's conceptual history. Nevertheless, it is important to restate the true nature of the debate over spontaneous generation, especially in view of later developments regarding the origin-of-life problem. Contrary to the accepted wisdom proclaimed in many historical accounts, the opponents of spontaneous generation during the second half of the nineteenth century did not represent the victory of the scientific spirit over its enemies. Evaluated within their own intellectual context, they embodied the religious, anti-materialistic, and anti-evolutionary position.

5

BETWEEN PASTEUR
AND DARWIN—
A DEAD END

Most scientists in the last decades of the nineteenth century considered Pasteur's experiments as decisive evidence against spontaneous generation of microorganisms. Despite attempts by several researchers to maintain the controversy, the perennial belief in spontaneous generation was abandoned at last. A radically different approach to the origin of life surfaced, based on the theory of evolution formulated in 1859 by Charles Darwin in *The Origin of Species*. From that time on, attention was focused on the emergence of an original living entity on the primordial Earth.

DARWIN AND THE ORIGIN OF LIFE

Darwin's theory represented the culmination of intellectual developments that began long before the nineteenth century. Challenges to the static conception of the universe and the Earth were raised in the seventeenth century, and belief in the fixity of species, each created separately by God, had begun to lose its hold in the eighteenth century. Historical evaluation of the living world gathered force in the first half of the nineteenth century following scientific discoveries in fields as varied as geology, paleontology, comparative anatomy, comparative embryology, and the geographic distribution of the different species (Mayr 1982:309–393; Bowler 1984:23–141). Indeed, in his book Darwin summed up enormous amounts of data from many different scientific disciplines, using them to validate his argument that during the history of the Earth new species were generated by descent from previous ones. Though this argument posed a serious threat to the traditional view of the fixity of species, the revolutionary significance of Darwin's theory stemmed mainly from its attempt to explain the origin of species through natural, material mechanisms, with no recourse to a supernatural, purposive power. Natural selection, Darwin's major explanation of the phenomena of evolution, results from processes that occur within populations of organisms, and is an inevitable outcome of basic facts of nature: the variability among organisms in a population, the ability of organisms to reproduce and transmit their properties from generation to generation, the fact that some of these properties are advantageous in various environments and the competition for resources among organisms within each population. On the basis of these facts, Darwin explained how new forms of life evolved in the course of history. Moreover, the Darwinian theory describes the development of all forms of life as an "evolutionary tree" that originated from the root of a common ancestor. Darwin wished to account for the evolution of "endless forms most beautiful and most wonderful" from "so simple a beginning" (Darwin 1963 [1859]:470).

The Darwinian theory assumed a starting point of the evolutionary process, a common origin—a single form or a few original forms from which all the species eventually evolved. The question thus arose, How did these original forms themselves arise on the primordial Earth? However, Darwin did not discuss this question in *The Origin of Species*. Both in the book and on other occasions, he said that he had "nothing to do with the origin of life itself" (Darwin 1963 [1859]:221), and at the end of the book, in a passage departing from the general spirit of the work, he spoke about "several powers having been originally breathed by the Creator into a few forms or into one" (Darwin 1963 [1859]:470). In a letter written several years later, Darwin expressed his regret that he had given in to public opinion and used the biblical term of the creation of life. In fact, he added, his intention was to refer to the origin of life as "some wholly unknown process." On the same occasion, he declared that "it is mere rubbish, thinking at present of the origin of life" (Darwin 1995:257). In another letter, written to the renowned geologist Charles Lyell, Darwin responded to the accusation that there was no point in discussing the evolution of species as long as the origin of life itself was unknown. The logic of this accusation, claimed Darwin, was similar to the argument that "it was no use in Newton showing the law of attraction of gravity and the consequent movement of the planets, because he could not show what the attraction of gravity is" (Darwin and Seward 1903:1, 140).

Thus, Darwin thought that, due to the state of contemporary scientific knowledge, there was no point in even thinking about the origin of life. Yet it may not come as a surprise that, as attested to by several of his letters, he did consider the subject. In a letter from 1871 he raised the hypothesis that a primitive form of life was produced on the ancient Earth as a result of various physical and chemical processes. It was a daring hypothesis at that time, and Darwin expressed his deep doubts about it in his famous expression, "But if—and oh! What a big if!" Nevertheless, he imagined "a warm little pond" that contained ammonia and phosphoric salts, which under the influence of light, heat, electricity, and so forth, formed "a protein compound . . . ready to undergo still more complex changes." (It should be noted that the term "protein"—in Greek, "primary matter"—was suggested earlier in the nineteenth century by the chemist Jacob Berzelius. At that time it was already known that proteins contain the chemical elements carbon, hydrogen, oxygen, and nitrogen, and that upon their decomposition several amino acids are produced.) In the same letter, commenting on the possibility of the present production of a primitive living system, given all the necessary conditions, Darwin said that the chance for such a process is nonexistent, because any organic compounds that could evolve into a living system would be "instantly devoured or absorbed" by some living creature. The origin of life had a chance only on the sterile, lifeless ancient Earth (Darwin 1969:3, 18).

The same argument, it should be pointed out, was raised independently by several biologists at the end of the nineteenth century and became widely known through the writings of the Russian biochemist Alexander Oparin, who in 1924 published a pioneering article on the origin of life (Oparin 1967:228). As we shall see, the breakthrough that launched twentieth-century research into the origin of life involved, among other things, specification of the particular conditions prevailing on the primitive Earth. These conditions, including the absence of any form of life, which were conducive to the synthesis of organic compounds, changed radically over time.

In the 1860s and 1870s, however, scientists interested in the question of the origin of life faced a difficult dilemma: Pasteur's experiments had demonstrated that even the most primitive forms of life could not be generated in organic solutions. Most scientists supported Pasteur's general conclusion that life can form only from previously existing life, and yet it was evident that a consistent theory of evolution logically implied that life had formed from inorganic matter through natural processes on the primordial Earth. The far-reaching materialistic implications of the theory of evolution led the major Darwinians of the age, among them T. H. Huxley, Ernst Haeckel, Carl Nägeli, John Tyndall, and August Weismann, to suggest different theoretical versions of abiogenesis, the generation of life from inorganic matter, supposed to have occurred on the early Earth (Kamminga 1988). A contributing factor to the formulation of these ideas was a change in attitude toward the living cell, culminating in the "protoplasmic theory of life."

THE PROTOPLASMIC THEORY OF LIFE

As we have seen, the cell theory, according to which all living tissues consist of cell units, was developed in the 1830s. The fathers of this theory, Matthias Schleiden and Theodor Schwann, emphasized the structural and functional importance of the cell nucleus and wall. In the 1850s, however, this emphasis was criticized by many biologists, who began to slight the role of the cell wall and to focus instead on the function of the "protoplasm," the half-liquid content of the cell, composed of proteinaceous material. The cell was conceived now as a lump of nucleated protoplasm, and appreciation of the protoplasm as responsible for the characteristics of life grew rapidly (Geison 1969). In a famous programmatic lecture delivered in 1868, T. H. Huxley described the protoplasm as "the physical basis of life." Life's main features, said Huxley, are manifested in a basic homogenous stuff made of protein, containing the elements carbon, hydrogen, oxygen, and nitrogen and identical in both animals and plants. The protoplasm is thus the clearest evidence for the unity of all life forms (Huxley 1869).

According to the protoplasmic theory, primitive organisms were nothing but naked lumps of protoplasm. This view enabled the Darwinists to apply their evolutionary outlook to the origin-of-life question. The German biologist Ernst Haeckel, Darwin's enthusiastic supporter, was quick to suggest that an ancient class of creatures, an intermediate stage between inanimate matter and a

living system, which he termed "Monera," indeed consisted of homogenous, structureless lumps of albuminous jelly, and was capable
of reproduction and nutrition (Haeckel 1866; 1902). It was probably not by chance that against this background, empirical evidence
seeming to confirm Haeckel's hypothesis soon suggested itself. In
1868 Thomas Huxley microscopically examined samples of mud
gathered from the ocean bottom a few years before. In these
samples Huxley found many lumps of gelatinous material, which
he viewed as remains of the primitive protoplasm of a member of
the ancient Monera (Gould 1982). In honor of Haeckel, Huxley
named the primordial organism *Bathybius haeckelii*. More evidence
for the existence of similar kinds of primitive organisms accumulated, and Huxley and Haeckel speculated that the sea bottom might
be covered with a layer of original protoplasm, the *Urschleim*—
evidence for the origin of life in the ancient past (Farley 1977:71–
77).

Haeckel's attitude, and that of other contemporary Darwinians,
to the question of the origin of life was first and foremost an expression of their worldview. Abiogenesis was a necessary logical
postulate within a consistent evolutionary conception that regarded
inanimate matter and life as stages of a single historical continuum
(Haeckel 1902:256–258). And yet no specific theory or mechanism
for the origin of life was suggested, and no statement of the specific conditions prevailing on the ancient Earth was made. The older
conceptual approach, based on belief in spontaneous generation,
had not yet been completely abandoned, and thus the suggestion
was made of a sort of spontaneous generation of the simplest organisms on the early Earth. A notable exception was the theory proposed by the German physiological chemist Eduard Pflüger.
Pflüger's hypothesis spoke of the formation of cyanogen compounds, which consist of carbon and nitrogen, in the hot temperatures prevailing on the primordial Earth. These compounds, he
claimed, underwent changes that produced more complex organic
compounds and finally led to the primitive protoplasm. Several
chemical difficulties were involved in Pflüger's scenario. Like other
versions of abiogenesis suggested around this time, it suffered from
the limitations of the protoplasmic theory, on which Pflüger relied (Kamminga 1988:6).

Toward the end of the nineteenth century, the protoplasmic
theory gradually lost its hold. New cytological studies revealed the
heterogeneity of the cell content and the crucial function of the

nucleus, which became associated with cell division. These findings strengthened traditional claims for the continuity of life and the formation of new cells exclusively from previous cells (Coleman 1965). With the rise at the beginning of the twentieth century of the new discipline of biochemistry, which focused mainly on the discovery and isolation of enzymes responsible for the catalysis of specific chemical reactions, the highly complex nature of the living cell became increasingly apparent (Kohler 1973:184–185). It had already been discovered in the 1870s that the famous *Bathybius* was only a chemical artifact, produced in the process of preservation of mud samples. Huxley had to admit his mistake, and the cause of abiogenesis suffered a severe blow (Geison 1969:284, n. 63). These developments served to highlight the dilemma faced by supporters of the Darwinian theory who wished to account for the origin of life. They realized that denial of the natural emergence of life from inanimate matter would strengthen the claim for the miraculous or divine creation of life. Yet the conceptual obstacles that prevented the formulation of a scientific theory explaining the origin of life were still numerous and formidable.

PANSPERMIA—ETERNAL LIFE IN THE UNIVERSE

Several renowned scientists at the turn of the century tried to solve the origin-of-life impasse by circumventing the very question of origin. As pointed out by the historian of science Harmke Kamminga, following Pasteur's conclusion that life can be generated only from previous life, it was suggested, mainly by physicists, that life had never emerged from matter but rather was eternal like matter itself (Kamminga 1982). Toward the end of the nineteenth century and at the beginning of the twentieth, several versions of a theory were suggested, according to which life had always existed somewhere in the universe where conditions were hospitable. These so-called panspermia theories, meaning "seeds everywhere" or "ubiquitous life," also assumed the ability of life, in the form of seeds or spores, to travel through space in various ways, and to somehow reach the Earth. All the versions of panspermia were based on the dualistic philosophical conception that life could not have arisen from inanimate matter because life and matter belong to two distinct categories. In addition, they relied on the prevalent evaluation of the laws of conservation of matter and force, which implied the eternity of the universe. In such a universe matter is eternal, and only form changes: new bodies are born, develop, and die, and some of them harbor life. Even those supporters of pansper- mia who were not explicitly committed to the eternity of the uni- verse and life preferred to "postpone" the riddle of the origin of life from planet to planet ad infinitum, and thus to relegate it, to- gether with the origin of the universe, to the sphere of metaphysi- cal questions outside the realm of science.

Among the panspermists were the English physicist Sir William Thomson (Lord Kelvin), the German physicist and physiologist Hermann von Helmholtz, and, at the beginning of this century, the Swedish physical chemist and Nobel laureate Svante Arrhenius. Though various considerations led the panspermists to formulate different versions of the theory, they all shared the view that biological organization was far too complex to have originated abio- genetically by random association of inorganic molecules. Thomson, for instance, thought it inconceivable that life could form from "dead matter." Life could proceed only from life (Thomson 1871). Helmholtz's dualistic position, and his claim for the eternity of life, are of particular interest. He was among the initiators of the antivitalistic trend in German physiology, who, starting in the mid- nineteenth century, wished to establish biology on a physicochemi- cal basis. Helmholtz's panspermia argument drew upon the conception of the organism developed by the German philosopher Immanuel Kant (Kant 1987 [1790], see, in particular, 248–261, 280– 283). This conception was indeed double-faced: On the one hand, due to the inherent purposive nature of biological organization, it is not possible to explain its emergence via materialistic mecha- nisms. Hence, Helmholtz assumed the eternity of life. On the other hand, once such an organization exists, it is possible to explain its development and activity using only physicochemical means, with no need for vital forces (Kamminga 1982:72–78).

In the eyes of its adherents, the theory of panspermia was re- garded as scientific, since the assumption of life's eternity automati- cally ruled out any notion of divine creation. Moreover, panspermia did not deny the validity of the Darwinian theory, which could account for the evolution of life after life was brought to Earth ready-made (Kamminga 1982:69).

The postulate of the eternity of life and its existence somewhere in the universe focused the attention of the panspermists on the physical question of the transfer of life to Earth. Thomson, Helmholtz, and others proposed that "seeds of life" had reached

Earth on meteorites. Ferdinand Cohn, discoverer of heat-resistant bacterial spores, suggested that spores could be carried by air currents into space and could sustain their viability for long periods of time despite the low temperatures. Arrhenius's panspermia theory included a detailed mechanism that was supposed to explain the interstellar transfer of life and its survival in space. Arrhenius relied on Clerk Maxwell's theory describing the weak pressure exerted by electromagnetic radiation, including solar radiation. Calculating the pressures needed for tiny bodies the size of spores to be pushed into space, he concluded that solar radiation could be responsible for such a transfer through long interplanetary distances. Ultraviolet radiation, Arrhenius thought, would not kill bacteria or bacterial spores, since its influence moderates in the absence of oxidative factors in empty space. By becoming attached to grains of interstellar dust, spores could land on a planet like Earth despite the pressure of the sun's radiation (Arrhenius 1909).

Experiments performed in the 1920s aimed at testing Arrhenius's claims, such as those carried out by the French agronomist Paul Becquerel (nephew of the physicist Henri Becquerel), showed that bacterial spores do withstand extreme cold for long periods of time, yet ultraviolet radiation kills them even in vacuum (Raulin-Cerceau et al. 1998:599). Moreover, there are other types of radiation in space, such as X-rays and γ-rays, that are highly dangerous for any bacteria or spores (Kamminga 1982:81). These findings, as well as the movement away from belief in the eternity of the universe following the rise of a new cosmology, led to the decline of the theory of panspermia among physicists. Most biologists did not regard these ideas as a serious solution to the origin-of-life question to begin with.

In the 1970s, ideas reminiscent of the notion of panspermia were raised anew by several scientists (Crick and Orgel 1973; Crick 1981; Hoyle and Wickramasinghe 1979a), and more recently, references to "panspermia theories" in the scientific literature have been quite abundant (Parsons 1996). It should be pointed out, though, that very different and conflicting notions are being grouped together under this category. Most versions of the theory of panspermia advanced at the end of the last century and the beginning of this one included two essential claims: the eternity of life in the universe and the transfer of life-forms to the Earth from outer space. There is no similarity whatsoever between these versions and more recent ideas, because no scientist today believes in the eternity of life in the universe. A few speculative theories for-mulated in the 1970s did share with past versions of panspermia the assumption that the emergence of biological organization was extremely improbable—almost a miracle. Extending the emergence-of-life arena to the entire universe and postulating the transfer of bacteria to Earth from outer space was suggested as a possible solution (Hoyle and Wickramasinghe 1981). However, these theo-ries are minority positions, not considered very seriously by most scientists.

Other recent findings raise the possibility of an import of or-ganic materials from outer space to the primordial Earth by mete- · orites and comets. It is the belief of many scientists that this exogenous delivery of organic compounds was the exclusive source of organics on Earth and a dominant factor in the emergence of life. Other scientists regard an exogenous source as an important addition to the synthesis of organic material on the early Earth (Whittet 1997). Yet obviously, this transfer to Earth of organic mol-ecules should not be confused with the notion of panspermia. It should also be added that, philosophically, there is no similarity between the historical versions of panspermia theory, which saw life and matter as two distinct categories, and the hypothesis re-cently proposed, following the discovery of putative microfossils in a meteorite from Mars, of the possible transfer of bacteria from Mars to Earth several billion years ago. I will deal with this subject in a later chapter. However, let it be said at this point that the ques-tion as to whether life emerged independently on Mars and on Earth, or whether there was some transfer of life-forms between planets in the solar system, is discussed within the evolutionary framework guiding research on the origin of life. This research as-sumes that life emerged very rapidly wherever environmental con-ditions were ready. Attention is focused now on the conditions that might have led to the emergence of life not only on the early Earth, but also on Mars or on any other celestial body.

THE ORIGIN OF LIFE—AN UNTOUCHED TABOO

After this short anachronistic digression, we move back again to the turn of the century in order to sum up the scientific and philo-sophical situation. The doubt cast on theories of abiogenesis fol-lowing the decline of the protoplasmic theory of life and the rejection of dualistic ideas about panspermia added to the uncertainty and confusion among scientists and philosophers regarding the relationship between matter and life. Among the expressions of this conceptual climate at the turn of the century was the rise of a neovitalistic trend that purported to explain biological phenomena on the basis of a nonmaterial "life force." Such vitalistic ideas, represented at the beginning of the twentieth century by the German embryologist Hans Driesch (Driesch 1914) and the French philosopher Henri Bergson (Bergson 1911 [1907]), expressed most strongly the dualistic conception associated also with the panspermia theory. The rise of neovitalism was a reaction to the tendency toward a mechanistic biology at the end of the nineteenth century. The supporters of this tendency, among them the famous German biologist Jacques Loeb, who worked in the United States at the beginning of the century, based all biological explanations on physicochemical mechanisms, often ignoring the unique characteristics of biological organization, especially self-regulation and teleological phenomena in embryonic development (Allen 1975).

As a logical extension of his mechanistic philosophy, Loeb postulated that in the foreseeable future life will be artificially produced in the laboratory (Loeb 1964 [1912]:7). Hans Driesch, on the other hand, claimed that control of the life force is transferred from generation to generation in a mysterious way, and that it is beyond our power to answer the question of the origin of life (Driesch 1914:38). The dualistic attitude of the neovitalists was noticeable also in Bergson's comment that the original life-forms were "tiny masses of . . . protoplasm" whose development was made possible through a "tremendous internal push"—a nonmaterial drive that overcame the resistance of inert matter (Bergson 1911 [1907]:99).

In still another intellectual development, an "organismic movement" was gathering force in biology during the first decades of the twentieth century. Distinct from both extreme mechanism and neovitalism, this movement drew its philosophical and scientific outlook from new research in physiology that concentrated on mechanisms of self-regulation and dynamic equilibrium manifested in the major organic systems of the body: the digestive, nervous, and blood systems. The interaction among the different components of the organism and the holistic nature of biological systems were the focus of this point of view (Fry 1996). Organismic biologists such as J. S. Haldane and L. J. Henderson rejected the vitalistic approach to the origin-of-life question, yet could not adopt the mechanistic option. Conceiving of biological organization as a functional complex system that possesses efficient mechanisms of self-regulation was not compatible with the claim that the emergence of life was based on the chance-like association of simple molecules. The question of the origin of life was thus also left unanswered by the organismic biologists (Fry 1996:177–188).

All the factors examined above—Pasteur's experiments, the rise and fall of protoplasmic and panspermia theories, the inability of both simplistic mechanism and vitalism to explain the origin of life, and the establishment of the field of biochemistry, which at the beginning of the century was already discovering the enormous complexity of the living cell—all these factors together turned the problem of the emergence of life into a total mystery. Many scientists during the first decades of the twentieth century preferred to ignore the subject. The renowned geneticist Herman Muller, in his review of this period, expressed the general feeling that "the subject of life's origin is so taboo" that it should be left untouched (Muller 1966:494).

6

THE OPARIN-HALDANE
HYPOTHESIS

Toward the middle of the twentieth century, the question of the origin of life became a legitimate scientific problem, open to research. What caused a change in the attitude of scientists to this subject, which at the end of the nineteenth century and during the first decades of the twentieth was regarded by many as an untouchable taboo?

Though actual scientific research on the origin of life did not begin until the 1950s, by the 1920s a few developments of significance to this field had taken place. The main contribution was made by two independently conceived papers, both titled "The Origin of Life." The first was published in 1924, in the Soviet Union, by the biochemist Alexander I. Oparin (Oparin 1967), the second in England, in 1929, by the biochemist and geneticist J. B. S. Haldane (Haldane 1967). While the long-term influence of Oparin's research was far greater than that of Haldane's, nevertheless, from a historical perspective, the publication of the two articles can be seen as a turning point. Both researchers claimed that a necessary step on the way to life was the abundant synthesis of organic compounds on the primordial Earth. They proposed, for the first time, specific hypotheses about the geophysical conditions on the ancient Earth and the constituents of the early atmosphere that made this synthesis possible. Oparin, and to a lesser degree Haldane, described certain processes of chemical evolution that might have led to the synthesis of more-complex organic substances, which gradually accumulated in the seas and oceans and created the "primordial soup."

Though Haldane's paper was incorporated into a biology textbook a few years later (Wells et al. 1934:650–653), according to the physical chemist and origin-of-life pioneer J. D. Bernal, his ideas were dismissed in England as "wild speculations" (Bernal 1967:251). Oparin's paper, on the other hand, was in harmony with the materialistic ideology at home, yet had little impact outside the Soviet Union. In 1936, Oparin published a book, *Origin of Life*, in which the hypothesis raised in 1924 was more thoroughly explored. The book was translated into English in 1938, but only after the Second World War did it receive wider attention from the international scientific community (Oparin 1953).

The major claims made by Oparin and Haldane are commonly referred to as the "Oparin-Haldane hypothesis." Beginning in the 1950s and continuing for several decades, the hypothesis served as the unifying framework for scientific research pertaining to the origin of life. It is thus crucial to examine the factors that made this breakthrough possible and to consider the contribution of each of the researchers to the formulation of this influential conception.

ALEXANDER OPARIN: ESTABLISHING THE "BIOCHEMICAL-METABOLIC TRADITION"

Oparin's main philosophical motivation in 1924 was antivitalistic. He attacked the arguments raised by supporters of the panspermia theories, who, pointing to the enormous complexity of biological organization, denied the possibility of the generation of life from matter. Oparin's objective was to show that all the major characteristics of life—the organization of the cell, metabolism, reproduction, and response to stimuli—have parallel manifestations in the inorganic realm and therefore can be explained by physical and chemical mechanisms (Oparin 1967:208– 216). Toward 1936, Oparin's position changed. In his later work he placed emphasis on the unique features of living systems and on the evolutionary development of these features (Oparin 1953:60–62). Though in 1924 there was still some resemblance between Oparin's extreme mechanistic views and several previous versions of abiogenesis, there is no doubt that by the 1930s his approach was entirely novel. Relying on new research fields developing at the beginning of the century—especially the rising field of biochemistry—and making use of new astronomical, geophysical, and chemical data, Oparin was able to formulate a specific, detailed scenario for the emergence of the first organisms.

In the last chapter I called attention to the connection between the impasse that the origin-of-life question had reached at the turn of the century and the decline of the protoplasmic theory of the cell. According to this theory, all the characteristic properties of life inhere in the protoplasm, the major constituent of the cell, which was perceived as a homogenous material, simple in both constitution and structure. The appearance of life on the ancient Earth was assumed to result from the association of inorganic compounds to form this simple substance. At the end of the nineteenth century, Haeckel and Huxley still entertained ideas about simple abiogenesis of lumps of protoplasm. However, with the rise of biochemistry at the beginning of the twentieth century, it was claimed that life can be manifested only at the level of the complex cell, which includes many enzymes and multifunctional inner structures. It was this conception that, for some time, created a seemingly impassable barrier between inanimate matter and living systems (Henderson 1970 [1913]:309–310; Kamminga 1991:103). Oparin and Haldane, while acknowledging the complexity of the living cell, suggested, each in his own way, a bridge between inorganic matter and life, pointing out intermediate steps that led to the evolution of the cell.

Oparin's idea for such a possible bridge was drawn from the young scientific field of colloid chemistry, which was coming into its own at the beginning of the century. In the 1860s, a group of chemical substances, including starch, glue, gelatin, and other proteins, was characterized according to its behavior in solution. The colloids (kolla—glue, in Greek) were distinguished from the crystalloids, such as sugar and salt, by their limited ability to dissolve and to crystallize, and by the cloudy suspensions they form when mixed with solvents. For several decades colloid chemistry was one of the major tools in the characterization of the cell. While it is clear today that behavior in solution is determined by the size of the solute molecules, this fact was not recognized in the early days of biochemistry. One of the main objectives of this new field was the chemical analysis of cellular constituents and protoplasm, analysis that was limited during these years by the resolving power of the optical microscope. In addition to the structures identified within the cell, the protoplasm appeared as a colloid—a kind of suspension of small particles in a liquid. It was claimed that this colloidal nature was responsible for the unique character of the cell. Most biochemical studies early in the century focused on the analysis of metabolic reactions and the identification of enzymes, the proteins that catalyze the different reactions within the cell. The enzymes identified at the time were defined as "colloidal catalysts" (Kohler 1973:185; Oparin 1953:173–175).

Only after the invention in the late 1920s of the ultracentrifuge, an instrument that facilitated the separation of very fine suspensions from solutions, and the invention of the electron microscope in the 1930s, did it become possible to analyze the cell content in molecular rather than colloidal terms (Jacob 1982:237).

Drawing on the chemistry of colloids, Oparin suggested various processes that supposedly occurred on the primordial Earth, leading to the emergence of primitive living systems. But first he had to account for the initial step necessary for life—the formation of high concentrations of organic compounds. In 1924, and more extensively in 1936, Oparin described the development of the solar system and the Earth, making use of astronomical theories then current and data concerning the chemical composition of the sun and other celestial bodies. He suggested a mechanism for the synthesis of organic from inorganic compounds based on a theory of the formation of petroleum from minerals developed at the end of the nineteenth century by the famous Russian chemist D. I. Mendeleev (Oparin 1967:217–226). Though Mendeleev's theory has since been rejected for a number of reasons, Oparin used some of its assumptions in a creative manner in devising his scenario. He provided evidence for Mendeleev's claim that in an early stage of the Earth's formation it already contained carbides—compounds of carbon and heavy metals, especially iron—and he described how these carbides were extruded from the depths of the Earth to the surface by volcanic eruptions. When meeting with steam in the atmosphere, the carbides formed hydrocarbons, the simplest compounds of carbon and hydrogen. Oparin pointed out that various celestial bodies, including comets and meteorites, contained carbides and hydrocarbons.

Ammonia, a compound containing nitrogen and hydrogen, was formed on the early Earth in a similar manner, and when it combined with hydrocarbons and other simple organic substances, more elaborate compounds developed. When temperatures dropped below 100°C, condensation occurred in the atmosphere, and consequently seas and oceans formed. In these bodies of water organic compounds—among them, chemical polymers similar to sugars and proteins—accumulated (Oparin 1953:108–136). Oparin (and also Haldane), like Darwin, called attention to the fact that these phenomena could not have happened unless the Earth was sterile at the time. Life can form from inanimate matter only prior to the existence of life. If life already existed, any newly formed organic compound would be immediately devoured by some microorganism, as we see in our present world (Oparin 1967:228; Haldane 1967:246). (It should be emphasized, however, that given a suitable environmental niche in which there were no heterotrophs—that is, biological systems dependent on external organic compounds for their existence—but only autotrophs—systems synthesizing their food from inorganic compounds—then the primordial scenario could have repeated itself [Wächtershäuser 1992:114].)

According to Oparin's hypothesis, organic compounds formed a sort of colloidal solution in the primordial ocean. From colloid chemistry it is known that when a solution containing certain polymers reaches a specific concentration, a process called coagulation occurs, in which two separated phases form. The more concentrated phase produces enclosed structures, which are separated from the dilute phase constituting the rest of the solution (Oparin 1967:228–230). The process of coagulation involves the association of polymers like sugars or proteins, under the influence of intermolecular forces, to form microscopic droplets, called "coacervates." Since coacervates are distinguished by their ability to absorb other substances from the surrounding solution, Oparin theorized that some of the coacervates in the primordial soup contained and absorbed from the outside small organic molecules, such as mono sugars and amino acids, as well as larger proteinlike molecules, which could have functioned as the first enzymes. Thus, a sort of primitive metabolism could have taken place within the droplets. Indeed, the ability of the coacervates to absorb organic compounds from the external solution and to maintain a set of metabolic reactions was demonstrated in several experiments conducted by Oparin and his group (Oparin 1953:148–160).

Upon absorption of materials, the droplets grow, and when they reach a certain size they divide in two. Oparin referred to this process as the reproduction of the coacervate units. As a result of this division, some of the properties of the "parent cell," especially the inner organization of the protein molecules responsible for the ability to absorb materials and grow, are transferred to the "daughter cells." Oparin raised the possibility that a sort of natural selection could function among the coacervates, in which droplets endowed with a more "successful" organization grow and reproduce faster than the less successful ones. He claimed that the ability to maintain a primitive metabolism, to grow and reproduce, and competition based on differences in adaptation to environmental conditions brought about the evolution of more complex and efficient systems (Oparin 1953:193–195). Finally, the stage was reached when all organic compounds in the environment were exhausted, and only those organisms that became autotrophic, devising ways to synthesize organic material on their own, survived and evolved. While the first creatures exploited the chemical energy stored in the organic substances in their environment, those that followed were forced to rely on alternative means to produce energy. The most efficient solution ultimately reached by the developing organisms was the tapping of solar energy in the process of photosynthesis (Oparin 1953:209–245).

It is important to note empirical and philosophical changes in Oparin's text of 1936 compared to the work of 1924. Following new astronomical and geochemical studies, Oparin reached the conclusion (realized by Haldane, for different reasons, in 1929), that the early atmosphere did not contain any oxygen, but was rather, to use the chemical term, a reducing atmosphere—rich in hydrogen-containing compounds, especially methane (CH_4, a carbon and hydrogen compound) and ammonia (NH_3, a nitrogen and hydrogen compound). This conclusion was informed by the discovery of methane and ammonia in the atmosphere of several planets in the solar system and by the resulting claim that such a chemical constitution is probably similar to the one that originally existed on the primordial Earth (Oparin 1953:64–104). The argument for a reducing atmosphere is highly important, because organic compounds form easily and accumulate under reducing conditions.

In his earlier text, Oparin was philosophically preoccupied with traditional materialism. Above all, he wished to characterize life in physical and chemical terms. He identified the "first organisms" with the products of the coagulation processes that occurred in colloidal solutions (Oparin 1967:229). By the 1930s, however, Oparin's philosophical and ideological positions were being influenced by the principles of dialectical materialism as formulated by Friedrich

Engels in his book *Dialectics of Nature* (Oparin 1953:31–33; Graham 1987). (Engels's book was never published during his lifetime, appearing for the first time in the Soviet Union in 1925.) Oparin applied to his origin-of-life theory the basic dialectical postulate that matter undergoes changes, evolving from one level of organization to the next, each level being characterized by specific new laws, including "biological laws" pertaining to living systems (Oparin 1953:162). In his 1936 book he emphasized the properties unique to life, especially complex organization and the purposeful nature of biological processes, viewing these properties as manifested at a certain historical stage of the development of matter. While competition among the coacervates, he now thought, stemmed from the different rates of growth caused by physical factors, the biological law of natural selection, an "all or none process," began to function only when the organic material in the environment was exhausted and the independence of the system, contingent on its internal organized enzymatic activity, was established. Oparin now perceived the emergence of life as an integral part of the general evolution of matter. In contrast to his thinking in 1924, he no longer assumed that crucial stages in the origin of life were the result of highly improbable chance events. On the contrary, he now claimed them to be a probable manifestation of the laws of nature (Oparin 1953:159, 249–251).

In his later publications, especially after the discovery of the structure of DNA in 1953 and the establishment of molecular biology, Oparin granted a substantial role in his origin-of-life theory to the development of nucleic acids and the genetic code. Yet even in these later works the emphasis is on the living system as a complex metabolic whole (Oparin 1968). In his view, such a metabolic system is a necessary condition for the development and functioning of the genetic material. It was Oparin who initiated the "protein-first" tradition that rejected the primary role allotted to nucleic acids in the emergence of life. This line of research denied the view that the organism is, first and foremost, a self-replicating genetic system (Oparin 1965).

J. B. S. HALDANE: ESTABLISHING THE "GENETIC TRADITION"

The achievements of biochemistry, the concepts of colloid chemistry, and the "petroleum theory" of Mendeleev served as a stimulus to Oparin in the formulation of his hypothesis of the origin of life.

Haldane was inspired by other sources: the rising field of genetics, the discovery of viruses, and the work of an English chemist, E. C. C. Baly of Liverpool, on the synthesis of sugars. In 1929 Haldane suggested that the early atmosphere contained little or no oxygen (Haldane 1967:246). He based this idea on the hypothesis that all the carbon now found in coal, in other organic remains, and in various mineral deposits was present in the past as carbon dioxide, tying up all the oxygen. Free oxygen could appear only after the emergence of the first plants and the beginning of photosynthesis. The primordial atmosphere, Haldane believed, contained carbon dioxide (CO_2), ammonia, and water vapors. It is important to remember that by 1936 Oparin was already talking about an atmosphere made up of methane, ammonia, free hydrogen, and water.

Haldane contended that because of the lack of oxygen in the early atmosphere, the layer of ozone (molecules containing three atoms of oxygen) was also absent, and unlike today, an uninterrupted penetration of active ultraviolet radiation from the sun was possible. This radiation, he pointed out, was highly energetic and could have aided in the synthesis of organic compounds from simple inorganic substrates. Haldane relied on the work of Baly, who applied ultraviolet radiation to a solution of carbon dioxide in water, resulting in the synthesis of sugars—organic polymers containing carbon, hydrogen, and oxygen (Baly et al. 1927). In the presence of inorganic substances containing nitrogen, Baly also succeeded in synthesizing amino acids, the building blocks of proteins (Baly et al. 1927:198.) Haldane thus assumed that with ultra-violet radiation, simple organic compounds could form in the atmosphere, followed by more complex substances, like sugars and proteins, which accumulated in the ancient oceans, constituting "a hot dilute soup" (Haldane 1967:246).

Oparin saw the colloidal coacervates as an intermediate stage between the inanimate and the living (Oparin 1953:136). From his biochemical point of view, he characterized life on the basis of its ability to perform metabolism using complex enzymatic systems. Haldane, on the other hand, raised the possibility, associated with the rising field of genetics, that "the first living or half-living things" were large organic molecules, capable of reproduction (Haldane 1967:247). Noting that the process of reproduction depended on the supply of a variety of molecules, he claimed that the primordial sea was a "vast chemical laboratory" producing the needed materials. There is no way to tell, he said, how long these living or half-living things remained in their primitive stage. However, at some point a sort of "oily film" was produced in the "vast laboratory," which engulfed the reproducing molecules and thus led to the emergence of the first cell.

Haldane's suggestion that the intermediate stage on the way to the first organism was a self-reproducing molecule was influenced, according to his own account, by the discovery of the viruses at the beginning of the twentieth century (Haldane 1967:243–245; Podolsky 1996:94). Much smaller than bacteria and causing many diseases in plants and animals, viruses could not be seen microscopically, but were detected by their actions on other cells. Haldane was impressed, in particular, by the bacteriophage, the "bacterium-eater," discovered by Félix D'Hérelle in 1917, which attacks a bacterium, reproducing itself inside the cell and eventually destroying it. The question as to whether the bacteriophage could be defined as a living creature was actively discussed in the first decades of the century (Podolsky 1996:83–97). It was claimed by some that a phage is capable of reproduction when the necessary building blocks are provided. Others pointed out that it can survive and reproduce only within a bacterium cell and at the expense of that cell. For Haldane it was the combination of these characteristics— the ability to reproduce dependent on supporting surroundings— that made the virus an ideal model for the intermediate stage between inanimate matter and a living system. He pointed to the analogy between the cellular environment of the viruses and the primordial soup that constituted a nourishing medium for emerging life (Haldane 1967:247).

Interestingly, Haldane supported the hypothesis proposed by the American geneticist Herman Muller, who compared the bacteriophage to a gene that copies itself within the cell (Haldane 1967: 245). According to Scott Podolsky, for Haldane the virus was more than a functional model; it was also the true phylogenetic "missing link" between inanimate matter and life (Podolsky 1996:94). Haldane indeed raised the possibility that life may have remained "in the virus stage for many millions of years before a suitable assemblage of elementary units was brought together in the first cell." (Haldane 1967:247).

Though identifying the first molecules on the way to life by their ability to reproduce, Haldane nevertheless emphasized the organic, dynamic nature of life and regarded the first reproducing molecules as only "half-living." In a truly living system, he contended, the function of any part, including genes, depends on the cooperation of all the other parts (Haldane 1967:245, 247). Haldane's evaluation of the nature of the first living system underwent many changes throughout his career (Podolsky 1996). In 1954 he expressed an Oparin-like position, claiming that the decisive step in the origin of life was the formation of a cell. The fact that even viruses are surrounded by a sort of membrane was cited by him as supporting evidence for his view (Haldane 1954). In 1963, on the other hand, at the Second International Conference on the Origin of Life, Haldane suggested that the first organism may have consisted of an RNA molecule that functioned as a single gene (Podolsky 1966:122). Historically, however, it is clear that Haldane's main contribution to the establishment of the origin-of-life field was his 1929 paper and its part in the famous Oparin-Haldane hypothesis.

LEONARD TROLAND: THE HYPOTHESIS OF THE "FIRST GENE"

During the second decade of the twentieth century, another noteworthy theoretical attempt was made to analyze the origin-of-life question on the basis of new concepts of genetics. Early thinking connected with the development of genetics focused on the rediscovery of Mendel's laws of heredity and involved the identification of the hereditary role of chromosomes in the cell nucleus. Following the studies carried out by the American geneticist T. H. Morgan and his group around 1910, a hypothesis was established concerning the material basis of the unit of heredity, the gene, and the location of different genes on the chromosomes was determined. At the same time, several theories explaining the origin of life in terms of the origin of the first gene were proposed (Podolsky 1996; Kamminga 1986:3–5). One of the most interesting was suggested by the American physicist and psychophysiologist Leonard Thompson Troland (Troland 1914; Troland 1916; Troland 1917). Troland's ideas received little attention upon publication, but they were adopted in part by the geneticist Herman Muller in the 1920s and 1930 (Muller 1966:495–498).

Troland's hypotheses about the origin of life formed part of a comprehensive theory devised to explain, with the help of a "single physico-chemical conception," all the "fundamental mysteries of vital behavior" on which vitalists rest their case (Troland 1914:102).

This conception was that of an "enzyme or organic catalyst," which Troland believed could solve the major problems of biology: the origin of life, the origin of organic variations, the ground of heredity, the mechanism of individual development, and the basis of physiological regulation (Troland 1914:92–93). For his highly ambitious project, Troland relied on discoveries of the many enzymes that control reactions in the cell and on new findings in genetics.

Troland's scenario for the origin of life assumed that in the primordial ocean suddenly appeared a primitive molecule endowed with catalytic ability, including the ability to catalyze its own formation and thus to multiply. In addition to the process of autocatalysis, this molecule was capable of heterocatalysis—catalyzing certain chemical reactions in its environment, which produced a protoplasmic envelope around itself (Troland 1914:102–104). In his early publications, Troland described this primitive molecule as an "enzyme," but later he identified it with the gene, the unit of heredity. He suggested that the "genetic enzymes" are made of nucleic acid (Troland 1917:342). It should be pointed out that until the 1940s, the chemical nature of the hereditary material was not known for certain, and many supposed that it is a protein, or a mixture of protein and nucleic acid. In 1917, Troland contended that only by supposing "that the actual Mendelian factors are enzymes" could the difficult enigmas associated with heredity be solved (Troland 1917:328). In attributing to the primordial gene the capability of both autocatalysis and heterocatalysis, he demonstrated an impressive insight, foreshadowing knowledge of the structure and function of genetic material that became available only in the 1950s and later. Moreover, based on a logical analysis of the origin-of-life situation (Troland 1914:104), he postulated the need to combine the functioning of autocatalysis and heterocatalysis, of a gene and an enzyme, in one molecule. This claim can be seen as an early version of the RNA-world idea suggested in the 1960s and supposedly vindicated recently through the discovery of ribozymes, enzymatic RNA molecules.

Attention should be drawn to some of Troland's basic assumptions, which would later appear in other "gene-first" models of the emergence of life. The gene was thought to form by suddenly following a highly improbable fortuitous "first event." Troland dealt with this situation by claiming that in the vast amounts of time during which life could have emerged, even a highly improbable event would become possible. Moreover, all that was needed was the production of a single autocatalytic molecule (Troland 1914: 105), a fortunate collision of molecules, or one "lucky accident," to borrow a phrase frequently used later. Following autocatalysis, out of this single molecule many copies would form. Troland predicted that in a similar fashion to the production of the first "enzyme," other fortuitous events could bring about the synthesis of other enzymes or genes. These, together with the products of the imperfect process of autocatalytic duplication of the first enzyme, constituted mutations or variations on the original theme, which would copy themselves as well and would produce other chemical substances in their environment. Natural selection among these various enzymes would result in the selection of the best products under the circumstances. Thus, the evolution of the first living systems would be secured (Troland 1914:110–112).

A PHILOSOPHICAL BREAKTHROUGH

Oparin's biochemical line of research characterized life first and foremost as a multimolecular, multi-functional metabolic system. Haldane's and Troland's genetic theories focused on reproduction and identified it with molecular self-replication. These different orientations did not reach full expression in the 1920s and 1930s, but later they became the two main lines of research in the origin-of-life field, and the conflict between their supporters, especially with the establishment of molecular biology, grew in intensity. The "gene conception," according to which the emergence of life coincided with the emergence of a self-replicating molecule, became the dominant one in the field. It should be pointed out, though, that the development of both the metabolic and genetic conceptions depended on the general assumptions grouped under the Oparin-Haldane hypothesis. The hypothesis, as already indicated, claimed the existence of a primordial atmosphere rich in hydrogen compounds, and the action of various energy sources on the constituents of the atmosphere. Under these conditions, both the accumulation of organic compounds and their chemical evolution in the primordial soup were made possible. The influence of the hypothesis on many scientists in the following years was crucial, and its major tenets fostered scientific research on the origin of life.

Not only was the Oparin-Haldane hypothesis crucial as a framework for experimental studies, but its importance also lay in its

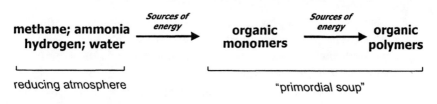

FIGURE 1. *The Oparin-Haldane hypothesis.*

Hydrogen-rich constituents of the primordial atmosphere were activated by electrical discharges, ultraviolet radiation, etc., to form organic monomers, such as mono sugars, amino acids, and nitrogenous bases. These building blocks, subjected to heat and the action of condensing agents, underwent polymerization to produce the biological polymers, such as polypeptides and polynucleotides. Monomers and polymers accumulated in the ocean to form the primordial soup, out of which the first living systems emerged.

philosophical significance. This consisted in positing a third alternative regarding the origin of life to the two previously entertained views. The hypothesis rejected the vitalistic option, according to which life and matter are two separate categories. At the same time, it denied the simplistic mechanistic outlook, which, not yet realizing the enormous complexity of the most simple living systems, equated life and matter. According to the materialistic-evolutionary alternative offered by the hypothesis, the passage between matter and life was continuous, yet unique properties of living systems arose that could be explained by natural processes.

More recently, following the discovery of new astronomical and geochemical data concerning conditions on the ancient Earth, several of the empirical assumptions of the hypothesis have been placed in doubt. The existence of a primordial reducing atmosphere seems rather uncertain. The likelihood of a primordial soup containing sufficient concentrations of organic material is questioned by some scientists. Such issues will be explored later in greater detail. In light of these doubts, there are many who call into question the whole Oparin-Haldane conception. Such contentions ignore the fact that the hypothesis is far from being a mere collection of empirical assumptions. Its philosophical content, the materialistic-evolutionary alternative, provides a necessary framework for scientific research on the origin of life despite the differences in opinion about specific historical scenarios.

From a current scientific point of view, research on the origin of life is an established fact, and the materialistic-evolutionary conception seems self-evident. It should be remembered, though, that the scientific-philosophical circumstances at the beginning of the century were quite different, and there was definitely a need for a conceptual breakthrough. It is the opinion of several authors, both historians of science and scientists, that that it was not by chance that those who contributed to the breakthrough were Marxists: Oparin, Haldane, the virologist N. W. Pirie, and the English physical chemist and crystallographer J. D. Bernal, who was among the first to suggest that chemical evolution did not occur in the "open sea" but rather on the surface of clay minerals (Graham 1987; Podolsky 1996:92, 107; Orgel 1973:14). These prominent scientists who contributed to the establishment of the study of the origin of life formulated a double-faceted philosophical position that spoke simultaneously of continuity and change. Their conception presented a materialistic view of nature, a view of continuity between matter and life, combined with the realization that new properties emerge that are unique to different levels of organization. Whether this philosophical conception was indeed an outgrowth of the principles of dialectical materialism is an open question. It clearly did foster a breakthrough, opening the way to the scientific investigation of the origin of life. It was claimed, for example by Muller, that Oparin adopted Marxist jargon in his later publications under the influence of political changes in the Soviet Union (Muller 1966:494). According to some historians, such as Loren R. Graham, the change in Oparin's position was determined conceptually and not because of political opportunism (Graham 1987:71, 90, 101). In his book on the origin of life, published in 1961, Oparin said, "Dialectical materialism makes it possible to accept the material basis of life without having to regard every phenomenon not included in physics or chemistry as vitalistic or super-natural" (Oparin 1961:5).

7

AN ERA OF
OPTIMISM

During the early decades of the twentieth century, different hypotheses for the mechanism of the emergence of life on the ancient Earth were suggested. Some of them, as already noted, identified the first living system with the "first gene." Yet as long as the chemical composition, structure, and modes of activity of genes remained unknown, it was Oparin's theory, rich in chemical and biochemical content, that served as the most convincing alternative. Indeed, several years after the publication in 1938 of the English version of Oparin's first book, biochemists started to conduct experiments guided by the Oparin-Haldane hypothesis.

STANLEY MILLER'S EXPERIMENT

The most significant of these experiments—the one that inaugurated the field of experimental prebiotic chemistry—was carried out in 1953 at the University of Chicago by Stanley L. Miller, who was at the time a graduate student in the laboratory of the chemist and Nobel laureate Harold C. Urey. Urey had proposed a theory concerning the origin of Earth and the other planets, presenting the hypothesis that the primordial atmosphere on Earth, similarly to the present atmosphere on several planets in the solar system, was rich in hydrogen-containing compounds (Urey 1952a). So it was not surprising that Urey and Miller wished to test Oparin's ideas to find out which chemical compounds could have been produced under the reducing conditions that prevailed on the early Earth.

Urey and Miller were interested in the composition of the primordial atmosphere for yet another reason. The biochemist Melvin Calvin at the University of California, Berkeley, who after the Second World War investigated the complex chemistry of photosynthesis (research for which he was awarded the Nobel Prize in 1961), had attempted in 1950 to synthesize organic compounds under prebiotic conditions. Calvin applied high-energy particle radiation, supposed to mimic cosmic rays or radioactive radiation within the Earth's rocks, to a mixture of carbon dioxide and water vapor, which simulated the primordial atmosphere. The experiment produced a very low yield of organic compounds (Garrison et al. 1951). Urey, suspecting that the reason for this was the oxygen-rich atmospheric content used by Calvin, suggested testing the behavior of a reducing mixture of gases instead (Urey 1952b:351). Miller built a sealed system of flasks and tubes in which water vapor was obtained by boiling an "ocean," a reservoir of water. The vapor was transferred to a gas mixture containing methane (a compound of carbon and hydrogen), ammonia (made up of nitrogen and hydrogen), and gaseous hydrogen, a mixture simulating the reducing primordial atmosphere. Electrical discharges served as an energy source, simulating the activity of lightning and causing the different gaseous molecules to interact. The products of this process were cooled and underwent condensation, and some of them dissolved in the simulated ocean. Miller repeated this series of steps for a week, during which the ocean changed from clean water to a red-brown solution containing various substances.

Urey and Miller's results created great excitement in the scientific community. Analysis of the compounds synthesized in the experiment clearly attested to the relevance of the Oparin-Haldane hypothesis to the origin-of-life question. Miller found that under reducing conditions about 10 percent of the available carbon was converted into organic compounds, about 2 percent of which were amino acids, the building blocks of proteins (Miller 1953). These results meant that amino acids, among the most important constituents of living material, could easily have been formed on the primitive Earth. Analyzing the chemical reactions that took place in the simulated atmosphere, Miller also described the intermediate stages in the synthesis of the major organic compounds obtained in the experiment.

Miller's experiment was the beginning of intensive research conducted in various laboratories around the world to evaluate the synthesis of organic molecules under different experimental conditions. Researchers used simple compounds consisting of the four basic chemical elements—carbon, hydrogen, oxygen, and nitrogen (and occasionally sulfur)—in an oxygen-free environment, applying to them various types of energy, such as electricity, heat, and ultraviolet radiation. In most of these experiments amino acids were produced, as well as the building blocks of sugars and occasionally those of nucleic acids (Miller and Orgel 1973). One of the most interesting findings was demonstrated by Juan Oró, then at the University of Houston, who succeeded in generating the nitrogenous base adenine—an important component of the nucleic acids DNA and RNA (Oró 1961).

We should pause here for a brief biochemical explanation. The major chemical constituents of every living cell are nucleic acids, proteins, sugars, and lipids. All of these are giant molecules built as polymers—long chains in which small units, the building blocks, are joined together in a certain order. The polymers responsible for the most important activities in the cell are proteins and nucleic acids. Proteins are made of amino acids, chemically connected to each other through the peptide bond. There are twenty different kinds of amino acids, and each protein is made of certain amino acids aligned in a specific order. The building blocks of nucleic acids are called nucleotides, to be described in the next chapter.

In addition to the support provided by the biologically relevant organic compounds produced in the experiments of Miller and other scientists, the Oparin-Haldane hypothesis and the theory of chemical evolution were further substantiated by the strong resemblance between the relative amounts of the amino acids synthesized in the simulation experiments and in proteins extracted from present-day cells. For instance, the most common amino acids in proteins, glycine and alanine, were the major products in prebiotic simulation experiments (Fox 1980; Eigen 1992). Moreover, the same result was obtained when the content of meteorites that have reached the Earth was analyzed. Most impressive were the relative amounts of organic compounds found in a meteorite that fell in 1969 in Murchison, Australia: the Murchison meteorite contained the same amino acids in the same relative quantities as the compounds synthesized by Miller (Miller 1987). Based on the knowledge that meteorites are a relic from the formation of the solar

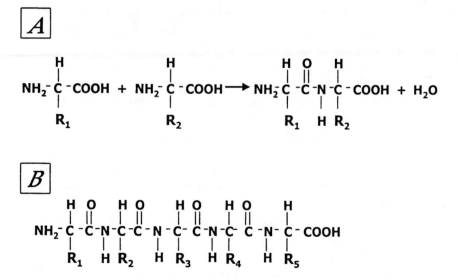

FIGURE 2. *Formation of a peptide bond and the structure of a polypeptide.*

A. *Formation of a peptide bond.* Two amino acids combine to form a peptide bond, –CO–NH–, by the elimination of the elements of water from the carboxyl group of one amino acid and the amino group of the next.

B. *A polypeptide chain.* Polypeptides are polymers made of amino acids linked to each other through peptide bonds.

system, their organic content could indicate that chemical processes leading to the synthesis of organic material were common in the early solar system, including the prebiotic Earth. The organic composition of meteoritic material was also regarded by many as pointing to the possibility of the emergence of life in the solar system at large.

The experimental successes during the 1950s and 1960s attracted physicists, chemists, and biochemists and led to the establishment of a legitimate scientific field devoted to prebiotic chemistry and research on the emergence of life. An important step in this direction was an international conference that convened in Moscow in 1957 dedicated to "the origin of life on the Earth," in which renowned scientists from many countries participated. Reports were given on the experimental synthesis of the various biological building blocks under prebiotic conditions, as well as on tentative results in the production of polymers resembling the major biological macromolecules (Sullivan 1970:104–120). Most of the studies in the 1950s followed the biochemical trend outlined by

Oparin, focusing on the prebiotic synthesis of amino acids and their association to proteinlike molecules. An important contributor to this line of research in the 1950s and 1960s who was instrumental in the development of the field of the origin of life was the American chemist Sidney W. Fox (1912–1998).

THE MODEL OF SIDNEY FOX: THE "PROTEIN-FIRST APPROACH"

In a long series of experiments performed over several years, Fox and his group, first at Florida State University and then at the University of Miami, suggested a model for the emergence of life that included the following stages: first, the spontaneous production of amino acids from the inorganic constituents of the primordial atmosphere, under the influence of high temperatures as an energy source (Fox and Windsor 1970); second, the condensation of amino acids, using heat as energy, to form protein-like polymers, called "proteinoids" by Fox (Fox and Harada 1958); third, the formation of cell-like spherical structures, called "microspheres," from a solution of proteinoids under specific physicochemical conditions (Fox 1984). Fox's synthesis of amino acids confirmed previous prebiotic simulation experiments: the relative quantities of the different amino acids were compatible with those known from regular proteins and from the analysis of meteoritic content. Fox referred, in this context, to a "thermodynamic order" reflecting the relative stability of the different amino acids (Fox 1980).

As for the synthesis of polymers, Fox had to face the challenge of forming peptide bonds between amino acids in an aqueous solution. This chemical process involves the removal of a water molecule from two adjacent amino acids, a difficult task to accomplish in aqueous solution. Under such conditions the strong tendency is toward hydrolysis—the disruption of the peptide bond by the addition of a water molecule and the liberation of free amino acids. In order to prevent hydrolysis of the peptide bond, either a condensing chemical agent must be used, which removes water by combining with it, or temperatures above the boiling point of water must be applied. As will be shown later, many researchers today achieve the formation of peptide bonds on the surface of various minerals. Fox applied the heating method, assuming that on the primitive Earth heat required for both the synthesis of amino acids and their polymerization could have been provided by hot lava or volcanic ash. He succeeded in producing peptide chains of lengths up to several hundred amino acids, resembling regular proteins in several properties. Unlike ordinary proteins, Fox's proteinoids contained various other building blocks in addition to amino acids. Furthermore, the chemical bonds between the monomeric units were occasionally different from the usual peptide bond. The most interesting property of the proteinoids, however, was their ability to act like enzymes—to catalyze different chemical reactions. Though this ability was rather weak in comparison to that of known enzymes, it was definitely significant (Fox and Dose 1977).

The next stage in Fox's model was achieved when concentrated solutions of proteinoids were heated to high temperatures and then slowly cooled. Under specific conditions of salt concentration and acidity, a large number of spherical particles—the microspheres—formed spontaneously. The microspheres in many ways resembled Oparin's coacervates, although they were characterized in terms of the biochemistry of the 1950s and 1960s rather than the chemistry of colloids. According to Fox's model, the enzymatic activity of the proteinoids organized within the microspheres enabled these structures to absorb additional proteinoids from the external solution and to grow and divide, thus forming a new generation of microspheres. Fox performed his experiments after the discovery of the role of DNA in cell-division processes and the identification of DNA as the carrier of genetic information. Nevertheless, like Oparin, Fox claimed that in the first stages of the emergence of life nucleic acids were not yet formed, and therefore genetic information was not yet relevant. Instead, he discussed the reproduction of information associated with the division of microspheres, referring to the transmission of the mode of organization and function of the proteinoids from generation to generation (Fox 1984).

Focusing on the production and function of proteinlike molecules and the formation of a primitive cellular system containing catalytic molecules as the primary factors in the emergence of life, Fox continued in the direction established by Oparin. This course differs radically from the genetic approach, according to which the primal event in the emergence of life was the appearance of a self-replicating molecule. The genetic approach also assumes that the synthesis of a system separated from its environment was a complicated and rather late event and that the first gene was "naked" in the primordial soup. Oparin and Fox, on the other hand, postulated the early and relatively simple appearance of a coacervate or a microsphere—a cell-like structure enveloped by a membrane

(De Duve 1991:201–202). The protein-first theorists described the next stages in the development of living systems, including the development of genetic material, as possible only in the framework of such an original cellular structure.

Fox's philosophical conception also echoed that of Oparin, posing the basic question, Was the emergence of life due to random events of very low probability, or did the first stages develop along certain nonrandom channels, each stage constraining and determining the one that followed it? Fox directed his attention to the physical and chemical factors in the prebiotic environment that were responsible for the selection of specific evolutionary pathways among all the theoretically possible chemical pathways. He emphasized the claim that the formation of proteinoids out of amino acids was statistically nonrandom. Starting the synthesis with a mixture of amino acids in equal concentration, he reported, the resulting polymer did not contain equal amounts of all the different amino acids; some of them were represented in a relatively higher proportion. Moreover, the sequence of amino acids in the polymer repeated itself in repeating syntheses. Fox thus asked, Why are the products not statistically random? Why are the different amino acids not equally represented? Why are they repeatedly arranged in a certain order? (Fox 1980; Kenyon and Steinman 1969).

All these facts, argued Fox, stem from processes of chemical selection or molecular selection based on the properties and structure of the various amino acids. There are small ones and large ones; some carry an electrical charge, positive or negative, and some are neutral; some are hydrophilic ("water-loving"), while some are hydrophobic ("water-hating"). These different chemical properties dictate a preferable chemical bonding between specific acids. In addition, the possible products differ according to their stability, and the chances of survival of the more stable polymers are obviously higher than those of the less stable. To summarize, due to chemical selection a certain order is formed in the polymer, determined by the properties of the reacting amino acids. In Fox's view, the ordered polymerization of amino acids was the strongest evidence of the tendency of certain prebiotic compounds toward self-organization, attesting to the non-random character of the processes of prebiotic and chemical evolution (De Duve 1991:140–143). This character is even more evident from the enzymatic activity of the proteinoids and the relatively easy formation of microspheres out of proteinoid solutions.

Fox's contention concerning the molecular order characteristic of proteinoids was rejected by researchers who were unable to repeat his experiments. According to them, in distinction to ordinary proteins, in which the sequence of amino acids is fully determined by the information in the DNA, the synthesis of protein-like polymers in the emergence of life entirely followed the "law of chance" (Kok et al. 1988). The most illustrious representative of the "chance approach" was the French biologist and Nobel laureate Jacques Monod, whose ideas will be discussed later (Monod 1974). Even the supporters of Fox's point of view are aware of the limited power of chemical selection in bringing about order or self-organization in the first proteinlike polymers, compared to the ordered sequences of amino acids in present-day cells (Kenyon and Steinman 1969:263–264; Wicken 1987:48). Nevertheless, they argue that without the physico-chemical constraints active in the prebiotic environment and limiting the processes of chemical evolution, it is hard to imagine the emergence of life. Were all the possible processes and products of the same probability, then the chance of the organization of a biological system would have been infinitely small— in fact, a miracle. According to this position, the emergence of a genetic, self-replicating system and of the mechanism of natural selection depended on previous processes of chemical and physical selection limiting the scope of molecular possibilities (De Duve 1991:211–217). These questions of principle will receive more detailed treatment in the following chapters.

We have noted that Fox's model was a continuation of the research direction set by Oparin, which focuses on the emergence of a functional enzymatic system based on proteins. By the 1960s, following the discovery of the structure of DNA, the genetic code, and the mechanisms responsible for the synthesis of proteins, movement away from the Oparin tradition grew stronger. The conflict between metabolic and genetic conceptions, which in the 1920s and 1930s was only hinted at, now became fully apparent. The strongest theoretical criticism raised by the supporters of the genetic approach against Fox, and in retrospect against Oparin, is that their models lead to an evolutionary dead end. The Darwinian mechanism that guarantees the development of new life-forms adapted to their environment requires continuity between generations— a mechanism for the accurate transmission of information from parents to offspring. However, this transmission should not be completely accurate: new variations must continuously arise so that natural selection can single out those most suitable to the environment. Oparin's and Fox's opponents argue that only a molecular mechanism of self-replication similar to the self-replication of nucleic acids, which produces mutants differing in the fidelity and rate of replication, can result in the evolution of structures and functions. Such evolution cannot be achieved, they say, in systems containing only proteins and no nucleic acids (Eigen 1971:498–503; Eigen 1992:13–16). The mechanism postulated by Oparin and Fox, which depends on the division of microspheres and the transmission of properties from parents to progeny, allows for a very limited inheritance (Maynard Smith and Szathmáry 1995:71–72). In addition, there is no dependable mechanism for the production of variations, which form the basis for natural selection.

Another criticism of Oparin and Fox, empirical in nature, challenges the biological relevance of their findings. It is argued that coacervates and microspheres could not have functioned as primitive "protocells" prior to the appearance of the first cell. The morphological resemblance between the microspheres and cell-like structures, claim the critics, is only external. Microspheres are generated by physical forces, and their growth, based on the absorption of materials from the outside, is entirely different from the processes of biological growth. Moreover, these spherical particles are produced under very strict environmental conditions and tend to decompose easily. Thus it is hard to believe that such specific conditions prevailed and lasted on the primordial Earth. The formation of these protocells also depends on a high concentration of the proteinoids in solution. Is it realistic to assume, ask the sceptics, that such high concentrations of polymers were present in the primordial soup? (Thaxton et al. 1984:170–176).

Oparin's and Fox's positions have to be evaluated in the wider context of the ongoing conflict between "protein people" and "nucleic-acid people," between the supporters of metabolism and those of self-replication. Some of the arguments raised by the two camps are empirical, relating to the possibility or impossibility of certain chemical processes on the primitive Earth. Even these conflicting empirical claims, however, rely on different philosophical conceptions adopted by the rival camps in their characterizations of the living system. I have already noted the differing attitudes toward the question of probability in the evolutionary process. In the following chapters, while discussing current theories of the origin of life, I will dwell in more detail on the controversies between the two camps. I will also point to recent developments in the study of the origin of life that indicate areas of convergence between the two traditions.

At this stage in our historical survey of the development of the scientific field devoted to the origin of life, it is important to emphasize that, though major parts of Fox's theory were later challenged by many researchers, his influence at the time was instrumental in turning the problem of the origin of life into a scientific subject. Though the relevance of his microspheres to the process of emergence is dismissed by many, this is not the case as far as the proteinoids are concerned. As will be seen later, various scenarios, metabolic as well as well as genetic, rely on the possibility of the prebiotic formation of proteinlike polymers possessing enzymatic activity as a crucial step in the emergence of life (De Duve 1991:140–143; Eigen 1992:31–32). Fox's philosophical contribution to the subject is no less important than his empirical contribution. Against the chance approach, Fox helped formulate the philosophical anti-chance conception, pointing to the role of strong constraints channeling the emergence of life and its evolution.

8

EVOLUTION IN
A TEST TUBE

In the ground-breaking experiment conducted by Stanley Miller in 1953, as we have seen, various organic compounds, including amino acids, were synthesized from a mixture of hydrogen-rich gases mimicking the primitive atmosphere. This experiment marked the establishment of the field of prebiotic chemistry and fostered the development of research devoted to the origin of life. Miller's experiment was followed by many others in which the building blocks of the major biological macromolecules—proteins, sugars, and nucleic acids—were synthesized under what were assumed to be prebiotic conditions. Most research at that time focused on the synthesis of amino acids, the polymerization of amino acids into protein-like products, and the search for conditions under which these polymers would organize into protocells capable of sustaining a primitive metabolism. Based on the many experimental achievements of the 1950s and 1960s, scientists entertained the optimistic belief that the origin-of-life question was heading toward a solution (Shklovskii and Sagan 1966:229–239; Orgel 1973:232).

WATSON AND CRICK: THE DISCOVERY OF THE DOUBLE HELIX

The year 1953 was crucial for the development of research on the origin of life not only because of Miller's breakthrough experiment. It was also in that year that James Watson and Francis Crick

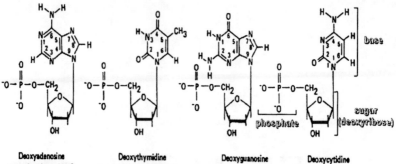

Deoxyadenosine monophosphate, dAMP

Deoxythymidine monophosphate, dTMP

Deoxyguanosine monophosphate, dGMP

Deoxycytidine monophosphate, dCMP

A. (*Above*) *Structure of mononucleotides.* Nucleotides are the building blocks of DNA and RNA strands. Each mononucleotide has three components: a nitrogenous *base,* a *sugar* molecule, and a *phosphate* group. DNA nucleotides contain the sugar deoxyribose and the four bases adenine, thymine, cytosine, and guanine. RNA nucleotides, not shown in the figure, contain the sugar ribose and the bases adenine, uracil, cytosine, and guanine.

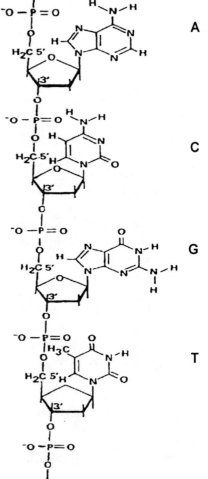

B. (*Right*) The *structure of a polynucleotide chain.* Nucleotides are linked to each other through phosphodiester bonds, in which a *phosphate* group forms a bridge between two nucleotides through the elimination of water from the *phosphate* group and two adjacent *sugar* moieties. Each polynucleotide chain is characterized by a specific sequence of *bases.*

FIGURE 3. The structure of nucleotides and their linkage to form a polynucleotide chain.

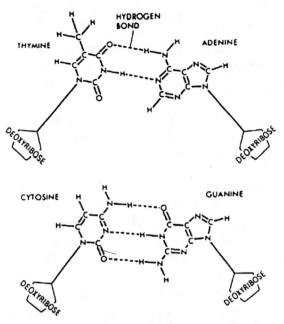

A. *Hydrogen bonding between bases.* Hydrogen bonds form between specific base pairs in the two strands of DNA. Adenine pairs with thymine, and guanine pairs with cytosine.

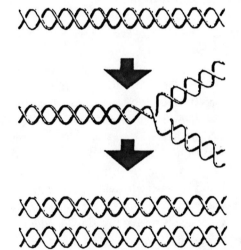

B. *Replication of DNA.* When the two helical stands start to separate, each strand serves as a template to which new complementary bases become attached through specific hydrogen bonding. The polymerization of the attached nucleotides to form two new strands results in the production of two helical molecules identical to the original helix. All of these processes are catalyzed by numerous enzymes.

FIGURE 4. *Hydrogen bonding between complementary bases and the replication of DNA.*

discovered the structure of nucleic acid, DNA, thereby establishing the field of molecular biology (Watson and Crick 1953a; 1953b). This discovery led to the understanding of the functioning of nucleic acids as hereditary material, responsible for the transfer of information from cell to cell and from generation to generation. During the following years, the mechanisms by which the information carried by nucleic acids directs the building of cell components, determining to a great extent cell functions, were unraveled. The accumulation of knowledge of the structure and function of nucleic acids at the molecular level had a major influence on the origin-of-life field. It became possible now to substantiate the "gene-first" theories of the past with detailed content and to design concrete experiments based on their hypotheses (Kamminga 1986:8; Podolsky 1996:109–110, 113, 117).

These new developments will be easier to follow after a brief description of the structure and function of nucleic acids and proteins. Every living cell (with a few rare exceptions) contains two kinds of nucleic acids, DNA and RNA, which are similar in composition and in some of their properties. Nucleic acids are polymers, whose monomeric subunits, called nucleotides, are linked to each other, forming long chains. Every nucleotide has three chemical components: a nitrogen-containing base, a sugar molecule, and a phosphate group. Every nucleic acid is characterized by a specific sequence of bases. There are four different types of nitrogen-containing bases in DNA nucleotides: adenine, thymine, cytosine, and guanine. In RNA, uracil takes the place of thymine. The sugar molecule in DNA is deoxyribose, and in RNA, ribose.

When Watson and Crick elucidated the spatial structure of DNA, it became clear that two strands of nucleic acid are paired via chemical bonds between their bases. These two strands coil around the same axis to form a double-spiral structure, the famous "double helix," which is similar to a spiral ladder or staircase whose "stairs" are formed by the bases of each chain. The double helix forms as the result of hydrogen bonding between specific pairs of bases along the two strands: because of their size and structure, adenine always forms hydrogen bonds with thymine, and guanine with cytosine. These specific pairs allow for the maximum possible number of hydrogen bonds and for maximum stability of the double helix. It is this complementarity between the specific bases that gives nucleic acids their unique properties and makes them carriers of information. Two daughter double-helical molecules are formed from each original double helix in the process of replication. Replication, which takes place in the nucleus during cell division, involves the separation of the two helical strands, each one serving as a template on which a new complementary strand is synthesized through the formation of complementary base pairs. In this manner, the specific sequence of bases in the DNA, which constitutes the genetic information, is accurately transferred from parents to offspring via the sex cells, as well as in regular cell divisions during the development of every organism. Replication, as well as all the other processes involving nucleic acids, depends on the activity of various biological catalysts, the enzymes, and on many other protein factors.

Because of their ability to form hydrogen bonds between complementary base pairs, nucleic acids, with the participation of appropriate enzymes, are responsible for other crucial processes in the cell. In transcription, segments of DNA are being transcribed into complementary segments of RNA. (Adenine in DNA is paired with uracil in RNA, and the RNA nucleotides contain ribose instead of deoxyribose.) The segments of RNA that carry a copy of the transcribed DNA in their complementary base sequence are called "messenger RNA." These messengers of information leave the cell nucleus and attach to structures called ribosomes located in the cytoplasm surrounding the nucleus. In a series of complex processes in which ribosomes, protein enzymes, messenger RNA, and other kinds of RNA take part, translation occurs between two chemical "languages"—the languages of nucleic acid and of proteins. "Words" made of triplets of nucleotides are translated into "words" of amino acids. The translation is carried out according to a detailed dictionary, the genetic code, which determines the strict rules of translation: specific combinations of three nucleotides code for specific amino acids. In fact, the translation from nucleotide triplets to amino acids is made possible again through the formation of base pairs in complementary nucleotide triplets on two different RNAs.

The translated amino acids are linked together to form all the proteins of the cell. These assembled proteins are responsible for the construction of the cell and for its varied activities: most importantly, for the constitution of all the enzymes, the catalysts that accelerate the chemical reactions of the cell. The function of every enzyme depends on its ability to combine with certain chemical substances about to undergo transformation. These substances bind to a specific area on the enzyme's surface, the binding site, which is part of the three-dimensional structure of the enzyme formed through a complex folding of the protein chain. Binding to the enzyme stabilizes the reacting substances and determines their spatial orientation so that their chemical association, breakdown, or transformation is accelerated at least a millionfold.

The three-dimensional structure of an enzyme is determined by its amino-acid sequence, and this sequence, as we have just seen, is determined by the sequence of bases in the DNA, mediated through a specific messenger RNA. The direction of the flow of information in every living cell is thus made clear: from DNA, through RNA, to proteins and cell activity. It is important to point out again that all the processes here described—DNA replication, its transcription to RNA, and the translation of the latter to protein—all consisting of many stages and intermediary products, would not be possible without the activity of enzymes and many other proteins. The fact that proteins coded by nucleic acids are needed for the production and activity of the very same nucleic acids is of great significance and will be discussed later.

The knowledge gained by molecular biology changed our understanding of cellular processes and opened up a new era in the investigation of evolution. As was emphasized by Darwin and by evolutionists ever since Darwin, the working of natural selection, the major mechanism of biological evolution, is contingent on the fact of variation. Every population of organisms, be they bacteria, elephants, or humans, consists of a variety of individuals. Those individuals that survive longer and leave more offspring in a given environment transmit their properties to the next generation to a greater extent than those that are less successful. This brings about gradual changes in the character of the population, which accumulate during long historical periods and produce entirely different organisms and eventually new species. The findings of molecular biology threw light on the source of variation within populations and explained the continuous appearance of new variants, the raw material for the process of natural selection and hence for the perpetuation of evolution.

The mechanism of DNA replication ensures the accurate transfer of genetic information on the basis of the complementarity of bases. While this process is generally trustworthy and accurate, it is not completely foolproof, and occasionally errors occur. Adenine complements thymine and guanine complements cytosine, but not in 100 percent of the cases. These errors of replication, as well as other processes that affect the genetic material, are called mutations, and the organisms that carry them are called mutants. Some mutations result in nucleic-acid sequences unable to undergo transcription or translation to functional proteins. There are many neutral cases in which mutations have no influence on the translated proteins and hence on cell function. On rare occasions, these errors grant the genetic system, and hence the organism, an advantage in its competition amongst organisms.

The evaluation of evolutionary processes in molecular terms led various researchers to the idea that processes of natural selection and evolution might occur not only in populations of organisms but also in populations of molecules (Orgel 1973:145–159; Eigen 1992:17–21). Starting in the 1960s and 1970s, the following question arose: Is it possible to describe processes of selection that will lead to evolution in a population of DNA or RNA chains? In more concrete terms, such selection processes require that given a mixture of polymers of different base sequences, a supply of nucleotides—the building blocks of self-replication—and the enzymes necessary for replication, polymers will compete with each other for "food," the building blocks, and each of them will replicate itself according to its ability. While replicating, these polymers will inevitably produce mutants that will also join in the competition. Each one of the competing sequences has properties of its own determined by its length, composition, and the specific order of its bases, and these properties determine its chances of success. This involves, for instance, different rates of replication, different stabilities of the nucleic-acid segments, and different fidelity of replication. Obviously, sequences that replicate faster and with fewer errors are better represented in the population. These ideas led to the first *in vitro* experiments in evolution—"evolution in a test tube" (Eigen et al. 1981; Eigen 1992:87–91).

SOL SPIEGELMAN: THE VIRUS MODEL

The first experiments to demonstrate Darwinian evolution in the laboratory were carried out in the 1960s by Sol Spiegelman and his colleagues at the University of Illinois (Spiegelman 1967). To test the idea in a relatively simple system, Spiegelman used a virus, Qß, that infects the gut bacterium *E. coli* and consists of only nucleic acid, RNA in this case, and a few proteins. In nature, the virus cannot reproduce independently. It has to insert its nucleic acid, the

RNA strand, inside the bacterium, and by using the genetic apparatus of the infected host to produce first an enzyme, Qß replicase,
which then catalyzes the replication of the viral RNA strand. It
should be pointed out that the process of replication occurring
within an infected *E. coli* cell consists of two stages: first, the viral
RNA serves as a template to which complementary bases are attached, forming a "negative" of the original strand. In the next
stage this "negative" strand serves as a template on which a complementary strand is synthesized. This "negative of the negative" is a
copy of the original viral RNA. The replication of a single-stranded
RNA clearly demonstrates the autocatalytic nature of the replication of nucleic acids. In the process of autocatalysis the reaction
product is identical to the starting material, and additional copies
of the original molecule are formed after repeated cycles. This is
what happens to the viral RNA within the bacterium cell: many
more copies of the virus are produced, which eventually cause the
bursting of the cell and its destruction.

Spiegelman started his experiment by isolating the viral genome, a short single strand of RNA, and its replicating enzyme, the
Qß replicase. He added to the reaction mixture the four types of
nucleotides required as building blocks for RNA synthesis and succeeded in demonstrating replication of the viral RNA outside the
bacterium cell (Mills et al. 1967). Using a series of sophisticated experimental steps, Spiegelman and his group then showed that when
conditions in the test tube are changed to simulate changes in environmental conditions, the viral RNA undergoes "natural selection" and acquires new properties. The mechanism of this *in vitro*
evolution is that because of many replication errors, the new copies of RNA produced in the system include mutants that possess
new characteristics. These acquired properties were revealed when
Spiegelman changed the environment in his test tubes, thereby applying new selection pressures (Safhill et al. 1970; Eigen et al.,
1981:82–83). In these changed environments, certain mutants were
fitter than others. First, it became clear that when the environment
"demanded" faster replication of the virus, shorter strands had the
advantage. This process culminated in an extremely short RNA seg-
ment, which was capable of replicating very rapidly but lost its ability to infect the bacterium. It turned out that parts of the viral
genome usually required for the infection of bacteria were lost under the pressure of the environmental conditions designed by
Spiegelman. In addition, when tested under high temperatures or

in the presence of an RNA-degrading enzyme, it was found that among the new mutants produced in the replication cycles were RNA strands resistant to high temperatures or to a degrading enzyme. These new RNA "species" apparently were able to form new three-dimensional structures, which were more resistant than the original strand.

Discussing the Oparin-Haldane hypothesis, I mentioned the analogy, suggested by Haldane in 1929, between a virus and an intermediary stage in the development of a biological system on the primitive Earth. The fact that a virus manifests the basic vital ability to reproduce through replication, combined with the virus's dependence on its host cell for propagation and survival, captured Haldane's imagination. He regarded the virus as an adequate model for the "half-living" systems on the primordial Earth that, he believed, were able to replicate but were completely dependent on the primordial soup for organic building blocks (Unlike the view of Haldane, who in 1929, like other contemporary biologists, considered the idea that viruses are a true phylogenetic "missing link" between inanimate matter and life [Podolsky 1996:94], the accepted view today differs. Viruses are seen as late-comers in evolution. They might have been part of the genetic system of certain cells but were later disconnected and set loose [Eigen 1993:42]. On the basis of such a history, viruses' amazing ability to exploit the genetic apparatus of various complex cells for their own needs can be better understood.)

In the 1960s and 1970s, following Spiegelman's experiments, the idea that viruses constitute a fitting model for emerging life was even more obvious than in 1929. Several scientists, first among them the German physical chemist and Nobel laureate Manfred Eigen, pointed out that Spiegelman's simple system of self-replication can serve as a good model system for experiments on the origin of life. Indeed, Eigen and his group at the Max Planck Institute for Biophysical Chemistry in Göttingen were among the first to apply the ideas about *in vitro* evolution to origin-of-life research.

Eigen and his colleagues found that under certain conditions evolution may be demonstrated even in a simpler system than the Qß one. Spiegelman's system included the replicating enzyme Qß replicase, the viral RNA as a template, and the nucleotides as building blocks. While conducting their experiments on the Qß system, Eigen and his colleagues were surprised to discover that synthesis of RNA and its further replication can be achieved even without adding an RNA template (Eigen et al. 1981:82–85). The mechanism involved when only the replicase enzyme and the four kinds of nucleotides are present was found to differ significantly from the mechanism of the template-induced RNA synthesis (Biebricher et al. 1981). The significance of these experiments for a hypothetical origin-of-life situation was, however, that an RNA strand can also be synthesized by the replicase from the nucleotide building blocks without an RNA template. Furthermore, processes of replication, mutation, natural selection, and evolution can be demonstrated in such a system.

LESLIE ORGEL: REPLICATION WITHOUT ENZYMES

Many origin-of-life researchers realized that despite the significance of the work of Spiegelman and Eigen, their respective systems were still too sophisticated to serve as a good model for a primordial Earth scenario. Viral replicase is a highly complex enzyme that is a product of later evolution, making it unrealistic to regard a system containing such an enzyme as mimicking the emergence of life. The question raised was whether processes of natural selection and evolution could be demonstrated under certain conditions even without enzymes. Experiments aimed at answering this question were carried on for many years in the laboratory of the English chemist Leslie Orgel at the Salk Institute in California. During the 1970s it seemed that the synthesis of complementary RNA strands on RNA templates according to the Watson-Crick pairing rules was indeed possible even without enzymes. Orgel synthesized oligonucleotides—short chains of linked nucleotides—and mixed them with free nucleotides. Instead of the complete Spiegelman system, which contained template, replicating enzyme, and building blocks, Orgel thus used only template and building blocks. A template made of uracil-bearing nucleotides directed the synthesis of a short strand of adenine-bearing nucleotides. An oligonucleotide whose bases were all cytosines directed the synthesis of an all-guanine oligonucleotide (Orgel 1994:60; Inoue and Orgel 1983).

In the previous chapter we noted the optimism felt by origin-of-life researchers in the wake of the prebiotic chemical experiments of Miller, Oró, and others, and the achievements of Sidney Fox in the production of catalytic proteinoids and microspheres. The promising experiments of Spiegelman, Eigen, and Orgel, which indicated the possibility of the production of a replicating system even with no enzymes, encouraged even higher hopes. However, experiments conducted by Orgel and others in later years revealed that the chances for non-enzymatic replication are very limited (Orgel 1994:60; Joyce and Orgel 1993:5–7). In many cases the short complementary strands produced on the template do not contain the "correct" chemical bonds between the nucleotides along the chain. Some of the "wrong" bonds hamper the continuation of the process. More important, it became evident that without the aid of an enzyme, even a correct complementary strand cannot serve as a template to complete the full cycle of replication. And yet in the 1970s and early 1980s, optimism still reigned, and the "gene-first" approach, based on the astounding developments in molecular biology, offered attractive models supposedly accounting for the emergence of life.

9

MANFRED EIGEN'S MODEL

Every living cell known to us is made up of several kinds of macromolecules, including nucleic acids and proteins. Nucleic acids—DNA and RNA—store and transmit genetic information, while the proteins perform enzymatic activity, which determines all the functions of the cell. The biological synthesis and activity of nucleic acids and proteins are totally interdependent: protein synthesis is directed by the information in nucleic acids—by the specific sequences of nitrogen-containing bases in DNA and RNA; nucleic acids are synthesized, replicated, transcribed, and translated into proteins only through catalysis by enzymes. The original emergence of this "vicious circle," which clearly demonstrates the involved nature of biological organization, is a cause for wonder among biologists. Proteins and nucleic acids are extremely complex molecules, a fact that makes it hard to imagine their simultaneous synthesis on the primordial Earth. And yet how could the one be produced without the other? This "chicken-and-egg" problem constitutes one of the major stumbling blocks in research on the origin of life. In fact the solution to the origin problem may be described as a resolution of the chicken-and-egg problem. Jacques Monod considered the probability of the original chemical production of such a vicious circle as "virtually zero"—as almost a miracle—and hence doubted whether a scientific solution to the origin-of-life problem is at all possible (Monod 1974:135–136).

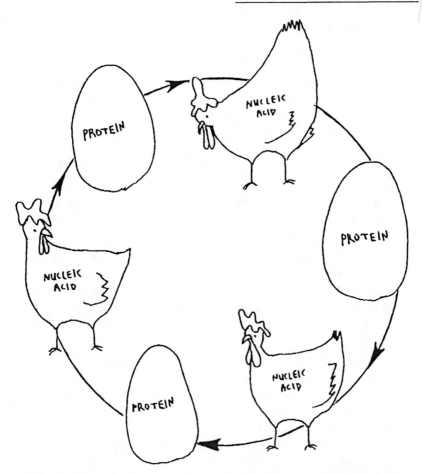

FIGURE 5. *Nucleic acids and proteins—the chicken-and-egg problem.* In extant cells, the synthesis and activity of nucleic acids and proteins are totally interdependent. These two types of macromolecules are too complex to have emerged simultaneously on the primordial Earth. How could the one have been produced without the other?

As discussed in previous chapters, most attempts to solve the chicken-and-egg problem assume that either the chicken or the egg—either proteinlike molecules or nucleotide-containing polymers—managed to appear first on the prebiotic scene and to get along for a while without the other. Sidney Fox, following Oparin's direction, described the first living system as based on the function of proteinoids that could have formed spontaneously. Fox claimed that microspheres produced out of a solution of proteinoids could have undergone processes of metabolism, growth, division, and evolution even without the presence of genetic material. The enzymatic activity of the proteinoids within the microspheres, according to Fox, led later to the synthesis of nucleic acids (Fox and Dose 1977). Orgel's experiments, following those of Spiegelman and Eigen, pointed to the other alternative: replication of genetic material, natural selection, and evolution without proteins (Inoue and Orgel 1983). Such a scenario, however, has to grapple with the question, How could replicating genetic material originally be formed? How did the first polymers capable of self-replication emerge?

Early hypotheses describing the "naked gene" assumed that such a polymer could have resulted from the random association of its building blocks, which were produced by random chemical processes in the primordial soup. It was thought that due to the extremely long stretches of time available for the emergence of life, even highly improbable events were possible (Troland 1914). Until the late 1940s, while the chemical nature of the genetic material was yet unknown, such hypotheses could be entertained (Alexander and Bridges 1926–1928:57). Even later, in the framework of the "optimistic era" characterized by belief in the power of prebiotic chemistry to work wonders (Eigen 1971:467; Eigen and Schuster 1978: 346), the "lucky accident" attitude could still be held (Crick 1981: 39). Recently, based on more realistic evaluations of prebiotic chemistry, the notion that self-replication could emerge by chance is rarely, if ever, entertained by origin-of-life researchers (Fry 1995). It is interesting to note that a current version of such a notion appears in Richard Dawkins's books *The Selfish Gene* and *The Blind Watchmaker*. According to Dawkins, whereas the evolutionary process is governed not by chance but by the mechanism of natural selection, which brings about the astounding organized complexity of living organisms and their adaptation to the environment, the process began with the chance formation of a "replicator," a molecule capable of self-replication (Dawkins 1978:16; 1986:140). Dawkins assumes long stretches of time during which life emerged, concluding that no miracle has to be invoked but rather an event of finite probability (Dawkins 1986:163). Though in *The Blind Watchmaker* Dawkins discusses in greater detail an origin-of-life theory suggested by the chemist Graham Cairns-Smith, he still considers "chance" as the only alternative to "design" as far as the origin of life goes. More will be said about this issue later.

THE DEVELOPMENT OF GENETIC INFORMATION—
THE "QUASI-SPECIES"

It is important to discuss Manfred Eigen's position on this subject because the theoretical and experimental work performed by Eigen and his colleagues in the 1970s and 1980s contributed significantly to the development of the genetic conception in the origin-of-life field. Eigen's theory concentrates on the stage at which a population of molecular sequences engaged in a primitive and highly inaccurate self-replication already existed (Eigen 1992:vi). Eigen and other researchers assumed that the earliest self-replicating sequences were of an RNA type. Eigen wished to analyze the development in such a population of what he defined as "genetic information": the emergence of specific RNA sequences—of specific composition, sequence, and chemical bonds between the units—capable of accurate and fast self-replication (Eigen 1992:126). That population, according to Eigen, can then develop into the next stage, which also involves protein synthesis.

In his first papers on the subject during the 1970s, Eigen dealt with the question of the synthesis of the first genetic polymers by relying on unspecified "random chemical events" (Eigen 1971:467). In his later work, though, he distanced himself from the "chance assumption," and he now puts greater emphasis on the deterministic aspects of origin-of-life processes. The synthesis of nucleotides, their linking into short RNA chains, and the beginning of replication could not have come about, says Eigen, without the participation of various catalysts present in the primordial soup, first and foremost, primitive enzymes. Fox's proteinoids, which could have functioned like primitive enzymes, undoubtedly preceded the genetic constituents (Eigen 1992:31–32). The previous claim about unspecified random events acquires a different meaning when, on the basis of Sidney Fox's findings, Eigen acknowledges the constraining role of physical and chemical selection in the formation of the first genetic polymers.

The disinclination to rely on chance is even more evident when Eigen deals with the question of the development of RNA sequences with a superior rate of replication, fidelity, and life span from the original population of short-lived, slow, and inaccurate polymers. Fast and faithful replication requires a certain length, composition, and order of the nitrogen-containing bases along the polymer, and therefore most polymers lacking these characteristics are totally unable to self-replicate.

Could the few "correct" polymers be synthesized by chance? What is the probability of the random appearance of a correct sequence? If we consider a single gene (or sequence) with only a thousand nucleotides and calculate the number of possible variants, assuming that each position in the polymer can be occupied by one of the four different nucleotides, the number that results is 4^{1000} or 10^{602}! Thus the probability of the random formation of a specific sequence out of all the possible ones equals one divided by this astronomical number, or 10^{-602}, the occurrence of which would amount to a miracle. In Eigen's words: "The volume of the entire universe . . . amounts to a 'mere' 10^{84} cubic centimeters. . . . The entire material content of the cosmos corresponds, weight for weight, to fewer than 10^{75} genes of the length assumed in this example." Or looking at our own planet: even if we assume the most favorable conditions for the synthesis of nucleic-acid sequences of this length, "after one thousand million years there would have arisen some 10^{50} fresh molecules" (Eigen 1992:10). The probability of the synthesis of a correct sequence out of all the possible combinations is not significantly different even if several variants and not just a single one were considered correct. We are still dealing with inconceivable numbers.

Seen in this light, correct sequences of nucleotides could not have arisen by a throw of the dice—that is, as the result of a chance event. Rather, the action of a principle of organization has to assumed, which limited, on the basis of certain criteria, the almost unlimited number of possibilities (Eigen 1992:11). Eigen claims that the principle of natural selection brought about the evolution of genetic information and hence, the evolution of a living system from inanimate matter (Eigen 1992:12–14). Against those who claim that natural selection could have started to act only once a living system existed, Eigen insists on the inverse option: only natural selection operating within an inanimate system could explain the emergence of life. Natural selection, he says, is a physical principle operating on physical systems in which certain conditions obtain. When sequences capable of primitive self-replication are present, and when a constant supply of chemical energy in the form of the necessary nucleotides is guaranteed, then natural selection is an inevitable outcome, leading to the emergence of an organized, information-bearing and -transmitting system (Küppers 1990; Lifson 1997).

Eigen's unique contribution to the analysis of self-replicating systems undergoing natural selection is based on his construction of theoretical models as well as on experiments with viral systems. This theoretical and experimental work resulted in the formulation of a new key concept in the origin-of-life field and in research on viruses, the "quasi-species" (Eigen 1992:79–80). Once a viral genome infects its host cell, it uses the genetic apparatus of the cell to replicate repeatedly and produce a population of replicated sequences. Since the viral replicating enzyme is inaccurate, this population includes many mutants. The "complex, self-perpetuating population of diverse, related entities," comprising the original viral genome and the distribution of its mutants, was called a viral quasi-species by Eigen. By using this term he pointed out that the population undergoes evolution as a whole, behaving like a biological species (Eigen 1993:42).

Examining the evolution of quasi-species, we should first note that self-replication of the RNA sequences is a competitive process: all the sequences compete for the same building blocks, and those mutants replicating faster and more accurately have a better chance of winning the competition and of being represented in greater numbers in the population. According to the traditional view of natural selection, the "wild type," or the "master sequence," within such a population would be the predominant sequence and the most successful one in terms of accuracy and rate of replication compared to all the other sequences. These less-fit sequences would consequently disappear from the scene. Based on theoretical and empirical studies, however, Eigen and his group concluded that the traditional picture is far from accurate (Eigen 1992:20, 82–86).

In the late 1960s, the Japanese theoretical geneticist Motoo Kimura formulated the "neutral theory" of molecular evolution. According to this theory, many mutants in each reproducing population are "neutral"; that is, they do not possess any advantage or disadvantage in comparison to the original wild type (Kimura 1983). This claim is highly significant for the analysis of the behavior of quasi-species. It turns out that many of the mutants produced in the repeating cycles of replication are indeed neutral—their sequences differ from the original RNA sequence because of erroneous copying—and yet this fact does not affect the efficiency of their replication. Thus, even "far" from the original sequence, despite the many changes in the original nucleotides that have occurred during copying cycles, peaks of fitness—of efficient replication—are achieved (Eigen 1992:27–30). Using new genetic techniques that determine the exact sequence of nucleotides and the percentage of each mutant in the population, Eigen and his colleagues have reached some highly interesting conclusions concerning the evolution of the quasi-species (Biebricher 1983; Domingo et al. 1978).

They found that the superior sequence in the population in terms of rate and fidelity of self-replication, the master sequence, is not at all master of the population, and its amount compared to all other sequences is minute. Many other mutants whose rate of replication is very close to the optimal rate under the given conditions "share the power" and contribute to the determination of the average sequence. Within Eigen's new theoretical framework, the wild type of the quasi-species is now defined as "the consensus sequence that represents an average for all the mutants, weighted to reflect their individual frequency" (Eigen 1993:45). It is this average sequence and not a particularly fit individual that determines the fitness of the quasi-species. Though it is produced following competition among its members, the quasi-species undergoes evolution as a whole. In contrast to the traditional picture, for a new quasi-species to replace the former one, it is not enough that a new superior sequence with a higher growth rate appears following a new mutation. For a stable quasi-species to be replaced by a new quasi-species, and for a new population to start to proliferate while the older population disappears, natural selection evaluates the contribution of all the sequences—the quasi-species as a whole. It is the fittest population that will have the upper hand (Eigen 1992:25–30).

Eigen's theoretical and experimental findings indicate that this joint contribution of all the sequences, based on the distribution of mutants in a quasi-species, leads to another significant result: the rapid evolutionary development of the population toward optimal self-replication. This conclusion conflicts with the traditional Darwinian wisdom, according to which there is no difference in the probability of "good" and "bad" mutations. Due to the inherent dynamics of a quasi-species and as a result of the quantitative relations within it, especially the high representation and unique distribution of mutants whose rate of replication is almost optimal, there arises in the population a preference for good mutations. The system seems to be moving toward the appearance of sequences that replicate more efficiently (Eigen 1992:29). This fact explains the astounding adaptation of many viral species to changing conditions while they still maintain their identity as species (Eigen

1993:42). The fact that populations constituting a quasi-species reach the stage of optimal replication relatively quickly can only be explained, Eigen claims, by the existence of some "guidance" toward superior mutants, which is provided by the unique distribution and dynamics of the developing quasi-species. The fast evolution typical of viruses could not have been achieved if these systems had to experience all the possible mutations in order to reach an advantageous one: in a random search, selection values rise and fall, and inferior mutants replace superior or neutral ones. In a quasi-species, on the other hand, the mechanism of selection guarantees a nonrandom and directed evolution (Eigen 1992:22–23).

Certain conditions are required for such a result: the length of the sequences in the population is restricted (10^3 to 10^6 nucleotides); the population numbers must be high enough (10^{10} to 10^{12} or more); the mutation rate must be high (though kept under a certain threshold, as will be shown shortly); and a far-from-equilibrium state has to be maintained through the perpetual supply of building blocks and energy (Eigen 1992:29, 31, 82).

Eigen claims that his quasi-species model provides an answer to the crucial question: How could genetic information have evolved on the primitive Earth? He also poses the historical question, How did this process in fact happen? His answer to the last question is by necessity speculative, and expresses his optimistic attitude. In his view, the primordial soup where life emerged constituted a chemically rich and plentiful environment that guaranteed, in a way similar to that of the host cell of a virus, far-from-equilibrium conditions. As in a replicating virus, the error rate was high because there was as yet no sophisticated enzyme to help the primitive replication process. The unique properties of the quasi-species enabled the first RNA sequences on the primitive Earth to evolve rapidly in a manner similar to that of fast-evolving viruses (Eigen 1992:31–36).

THE INFORMATION CRISIS AND THE "HYPERCYCLE"

The next stage in Eigen's model aimed to answer the question, How did RNA sequences evolve containing enough information not only for their self-replication but also for the construction of protein chains? The major change that had to take place for this to happen involved the lengthening of the replicating sequences so that they could store the additional coding information needed for proteins. Here, Eigen faced a serious dilemma: one of the necessary conditions for the existence and evolution of the quasi-species is the presence of short sequences. Since replication occurs without the participation of a sophisticated enzyme, many copying errors are produced. It is this fact that, among others, enables the quasi-species to develop rapidly toward optimal information. As long as the sequences are short, errors do not accumulate in each chain. Accumulation of excessive errors in successive replications could result in the destruction of the original information. Eigen calculated the "error threshold" that determines the maximal chain length under certain mutation rates that will still allow for the preservation of the original information. Under prebiotic conditions this probably amounted to about one hundred nucleotides (Eigen and Schuster 1977:555–557; Eigen et al. 1981:88–91). However, the synthesis of a protein requires longer RNA sequences, and longer sequences require a protein, that is, a replicating enzyme, in order to replicate under the error threshold. In short, a longer RNA could not be generated without a protein, and a protein could not be produced without a longer RNA (Eigen 1992:109). This dilemma, which Eigen called "the information crisis," was called by the English biologist John Maynard Smith "the catch-22" of the origin of life (Maynard Smith 1986:118).

According to Eigen, the only way to leap this hurdle was cooperation among short RNA sequences. The "goal" was to combine the information carried by single RNA segments into one system without crossing the error threshold. Had the cooperation been achieved through forming one long chain while no accurate replication enzyme yet existed, the problem would not have been resolved (Eigen 1992:109). Instead, Eigen postulated the emergence of a complex cycle: an ensemble in which several quasi-species units participated, each unit making it possible for the next one in the cycle to replicate. He assumed that at this stage each quasi-species unit, based on its limited amount of information, was able to produce a short protein that acted as a weak replicating enzyme. Such an enzyme then enhanced the replication of an adjacent quasi-species, which in turn produced an enzyme that catalyzed the replication of its own neighbor. The cycle could close when the last unit catalyzed the replication of the first one. This complex organization, which Eigen called a "hypercycle," made it possible to pull together the information contents of each one of the participating units (Eigen and Schuster 1977). It was subsequently suggested that hypercycles could have evolved containing RNA strands functioning themselves as enzymes, instead of having to produce protein enzymes (Eigen 1992:43). (The discovery of RNA enzymes, called ribozymes, will be discussed later.)

A short comment is in order here. "Cooperation" and similar terms used here do not imply any intention or purpose. Cooperation and other forms of organization manifested by systems undergoing evolution were the result of natural selection. The systems that did not adopt such helpful strategies had less chance of survival. The hypercycle organization is seen by Eigen as a step taken by emerging life that led to the evolution of living systems as they are known to us.

There is evidence, Eigen points out, of the existence of limited hypercycles in the life cycle of certain viruses: positive feedback follows when the protein produced by the viral genome, through the molecular apparatus of a host cell, enhances the replication of the viral genome itself (Eigen 1992:107–108; Eigen 1993:46). In order, however, to overcome the information crisis at the emergence of life, a more complex hypercycle would have been required, which would involve cooperation among several units, contributing to the development of positive feedback in the larger loop (Eigen 1992:109–111).

In order to secure interaction within the hypercycle, all the replicating units would have to be proximal. A sort of primitive cell must separate them from the external environment. Although in his earlier works Eigen emphasized the necessity of a compartment at this evolutionary stage, his focus was on the organization into a hypercycle (Eigen and Schuster 1977:564; 1978:368; Eigen et al. 1981:90). He also pointed out that in order to avoid having interfering boundaries, compartmentation must be a rather late event in the emergence-of-life process (Eigen et al. 1981:91–92). The emphasis is entirely different in his 1992 study, in which he speaks of a "compartmented hypercycle" (Eigen 1992:43), and claims that the integration into a hypercycle and its physical enclosure in a compartment had to be accomplished at the same time. Neither a hypercycle without a compartment nor a compartment that did not contain a hypercycle would have been sufficient (Eigen 1992:109–114). Both forms of organization together could have created a division among enclosed hypercycles, and in addition to cooperation within each "cell," competition could have developed among the cells. A process of natural selection would now take place among entities on a higher level of organization. Each mutation in one of the units within the cell that improved the activity of a certain enzyme would be manifested in a more efficient hypercycle and would have given it an advantage in comparison to other hypercycles. Following this process, more efficient enzymes could have gradually formed, enabling the progressive lengthening of the RNA sequences. Finally, the system would have been capable of synthesizing a replicating enzyme efficient enough to ensure the combination of RNA segments into one longer genome. At this stage, Eigen is talking about the emergence of the genetic apparatus as known to us, including the genetic code (Eigen 1992:43–45).

Eigen's model has been criticized for various reasons, both empirical and theoretical. According to several critics, among them the English physicist Freeman Dyson, Eigen imposes too strict a demand on the emergent system: On the one hand, any model based on self-replication has to assume quite an accurate replication. On the other, evolution through natural selection requires an error rate that will produce a wide enough spectrum of competing mutants. Dyson considers it unrealistic to expect emerging systems to navigate their way through these two conflicting demands (Dyson 1985:37). According to Dyson, early biological systems should be characterized by their tolerance of a high error rate. Such systems, he claims, were not engaged in self-replication but rather in metabolism of the kind assumed by Oparin (Dyson 1985:73). Another criticism that comes from Fox and other supporters of the "proteinic direction," who regard life not as a self-replicating unit but as a multimolecular functioning whole, points to an additional problem. Eigen's model assumes that following the transfer from the quasi-species stage to the hypercycle stage, protein synthesis starts somehow on the RNA strands. Here, they say, Eigen begs the question: the hypercycle is supposed to explain how protein synthesis started and not to presuppose such a synthesis (de Duve 1991:187). It seems that this criticism is answered by the ability of ribozymes, RNA strands themselves, to function as enzymes. Yet, as will be shown later, the emergence of enzymatic RNA is not less problematic than the synthesis of proteins in the system.

A more serious criticism in the same vein refers to the fact that, according to Eigen's model, RNA sequences in the quasi-species are supposed to self-replicate with growing efficiency. As was shown in Spiegelman's viral experiments, natural selection leads to the formation of increasingly shorter segments that replicate faster and faster. Thus there is no reason, say the critics, for the mutants in the quasi-species to start producing proteins. Jeffrey Wicken points out that information for making functional proteins is burdensome to bare replicators. On the basis of evolutionary logic, such a change in the behavior of RNA strands could occur only within a system that will benefit from it—for instance, within a microsphere whose existence and survival depend on the emergence of efficient enzymes (Wicken 1987:104). According to this line of argument, a more successful model for the emergence of life would describe the evolution of a quasi-species inside an already existing microsphere, combining the advantages of both the metabolic and genetic conceptions.

Several historical, empirical claims have also been raised against the optimism of Eigen's model, which assumes a chemically rich prebiotic environment producing all the necessary building blocks and polymers in abundance. Since Eigen focuses on the stage when RNA sequences already existed, he tends to ignore the difficulties on the way. Indeed, in the last few years, in light of new data in several scientific fields, the optimistic attitude toward prebiotic chemistry has been called into question. The various contentions against Eigen's model notwithstanding, his theoretical contribution to research on the origin of life, in particular to the question of the evolution of genetic information, is crucial. The concept of the RNA world, a dominant paradigm in the field, is greatly indebted to Eigen's work.

10

CRISIS—REAL
OR FICTITIOUS?

T he 1950s and the 1960s, and to some extent also the 1970s, were hopeful years for research on the origin of life. In the famous Urey-Miller experiment and in many other experiments that followed, synthesis of the major building blocks of every living system was achieved under conditions believed to prevail on the primordial Earth. Moreover, in some of these experiments, various biological polymers and even complex structures such as cell-like systems and populations of self-replicating molecules manifesting *in vitro* evolution were obtained. As noted in the previous chapter, Eigen's assumptions regarding a plentiful prebiotic chemical environment also belong to these years. The general belief in the field was that the Oparin-Haldane hypothesis had provided a framework within which the origin-of-life problem was heading toward its solution. Recently, following the presentation of new data and the suggestion of new models of the early Earth and solar system, ideas about prebiotic physicochemical conditions have changed, and the optimistic outlook has been replaced by a more cautious approach. The reevaluation of some of the previously accepted hypotheses has resulted in the suggestion of new scenarios for the origin of life on Earth.

It is not at all clear now whether the primordial atmosphere was reducing, that is, rich in hydrogen-containing gases. Hence the notion of an easy prebiotic synthesis of organic compounds under reducing conditions has been called into question. Despite the fact that this change involves empirical claims, it is nevertheless referred to in philosophical battles. Creationists, long-standing enemies of scientific research into the origin of life, rely on the new evaluation in order to speak of a "crisis" that science cannot resolve. They also revel in data indicating that the time available for the emergence of the first living systems was much shorter than previously thought. The natural emergence of complex biological organization already evident in the simplest cell, they claim, is even less likely within such a short geological time frame. They conclude that the need for a designer is strongly supported by the new findings (Thaxton et al. 1984:209–210). Scientists, adopting the opposite philosophical point of view, regard the same empirical data in a completely different manner. Instead of viewing the situation as a crisis, they consider it a challenge that calls for new ideas about the mechanisms responsible for the emergence of life under prebiotic conditions.

THE NATURE OF THE EARLY ATMOSPHERE

According to a central claim of Oparin and Haldane, the early atmosphere on Earth, unlike the present one, did not contain free oxygen. An atmosphere free of oxygen, they reasoned, would enable the synthesis and accumulation of organic molecules—the first step on the road to life. In 1936, Oparin argued for the existence of a primordial reducing atmosphere, containing a mixture of methane, ammonia, free hydrogen, and water vapor. Oparin based his ideas on theories first formulated during the 1930s dealing with the formation of the solar system and the Earth. He also relied on astronomical findings concerning the chemical constitution of the universe and the atmospheres on other planets within the solar system (Oparin 1953). It became clear that the atmospheres on the Jovian planets—Jupiter, Saturn, Uranus, and Neptune—contain hydrogen-based compounds, and the general assumption was that these atmospheres are the remnants of the primordial nebula that condensed to form the solar system. It was first proposed by H. C. Urey that since the Jovian planets have very large masses and their upper atmospheres are very cold, it is impossible for atoms to escape from their gravitational field. On the other hand, the terrestrial planets—Mercury, Venus, Earth, and Mars—which are smaller and closer to the sun, gradually lost their atmospheres to outer space, later acquiring a secondary atmosphere. It was thought for many years that this atmosphere was also reducing, and it was this notion that guided the Miller-Urey experiment (Urey 1952a).

Whether the Earth ever had a first "nebular" atmosphere is still debated today. It is commonly accepted, however, that the atmosphere in existence when life emerged formed as a result of volatile outgassing, the release of trapped volcanic gaseous compounds from the upper mantle of the planet. As elucidated by James J. Kasting at Pennsylvania State University and others, the composition of the emitted gases was determined by the internal structure of the Earth and the chemical composition of the upper mantle (Kasting 1993). According to the "cold homogeneous accretion model," which dominated thinking for many years, because of the presence of iron metal in the mantle, the nature of the emitted gases was reducing. More recent evidence, however, supports the "hot heterogenous accretion model," which hypothesizes that iron metal was removed from the mantle from the start and was concentrated in the Earth's core, creating an oxidized mantle. Thus the released volcanic gases forming the early atmosphere were mainly carbon dioxide, water, and nitrogen (Levine 1985), not methane, ammonia, and free hydrogen.

The question of the composition of the early atmosphere is still hotly debated. Several crucial factors seem to tip the scale toward a non-reducing atmosphere, but the picture is not at all clear, and there are indications that even though the major component was carbon dioxide, methane was present. Some scientists still insist on a reducing atmosphere, while others favor a nearly neutral oxidation state (Sagan and Chyba 1997:1217). This debate is crucial to the origin-of-life question, since the postulation of a non-reducing atmosphere makes it difficult to explain the generation of simple organic molecules and their chemical evolution to more complex ones. It should be emphasized that these debates are taking place within the framework of the now commonly accepted general "accretion model" of the formation of the planets in the solar system, according to which the Earth formed through a process of gradual accretion from lumps of material orbiting the young sun. This process involved the collision and sticking together of particles, gradually increasing in size from dust grains up to tiny planets, the so-called planetesimals (Gaffey 1997). The model is based on an idea suggested in the 1940s by the Russian geophysicist Otto Schmidt and revived following the Apollo space program in the

1960s. The craters revealed on the moon attested to the stormy early history of the solar system, when, as part of the accretion process, newly formed bodies collided with the young planets (Allègre and Schneider 1994).

Under the force of the continuous collisions and the impact of meteorites and other bodies on its surface, the inside of the Earth melted, giving rise to intense volcanic activity and the release of volatiles. As indicated, most researchers believe that carbon dioxide and water were the major volatiles released (Allègre and Schneider 1994:47). While rejection of the notion of a primordial reducing atmosphere casts doubt on the Oparin-Haldane scenario for the origin of life on Earth, the accretion theory might rescue this scenario. An important part of the theory is the continuation of collisions and impacts of meteorites and comets on the surface of the Earth for half a billion years after its formation. A widely accepted view considers the delivery to Earth of volatiles, water and organic compounds by the impacting bodies during the period of heavy bombardment as a major source of the building blocks of life, exclusively or in addition to endogenous synthesis (Whittet 1997; Irvine 1998).

The existence in space of masses of various organic compounds was recently established by extensive evidence (Tielens and Charnley 1997). In all probability, abiotic synthesis of such compounds took place in the solar nebula and occurs still in the interstellar medium out of which the solar nebula condensed. The amount of organic material delivered to the Earth today by meteorites and interplanetary dust is estimated to be around three hundred thousand kilograms per year. During the late heavy bombardment of the Earth, the rate might have reached fifty thousand tons per year. Such a rate could have produced the current total biomass in approximately ten million years! (Whittet 1997:257–258). According to another calculation, the number of comets required to deliver an amount of water sufficient to fill all the oceans on Earth is around one thousand—not an unlikely number during the period of bombardment (Owen and Bar-Nun 1995). Prebiotic materials delivered to the Earth included, among other organic molecules, the key compounds hydrogen cyanide and formaldehyde, in addition to several amino acids. I have already mentioned the meteorite that fell in 1969 near Murchison, Australia. The Murchison meteorite contained, among a rich variety of organic molecules, the same amino acids in the same relative quantities as those synthesized in the Miller-Urey experiment. Whereas in the

1970s these facts were seen as evidence of prebiotic processes that took place throughout the solar system, including the Earth, today they support the hypothesis of the exogenous delivery of organic material to the early Earth (Orgel 1994:56). Additional evidence is provided by observations of Halley's and Wilson's comets, which proved that comets are even richer in organic compounds than meteorites (Gaffey 1997:186–187; Irvine 1998:376; Allen and Wickramasinghe 1987; Huebner 1987).

Supporters of the hypothesis of the massive delivery of organic material to the primordial Earth by meteorites and comets have to face the problem of the enormous heat caused by the impact on Earth of the colliding bodies and the effect of this heat on the organic compounds. Calculations indicate that organic molecules would not survive impacts over a certain speed. This fact seriously limits the size of the accreting bodies capable of delivering prebiotic material (Whittet 1997:257). However, mechanisms have been envisioned that might have assured delivery, such as the protection of organic molecules within comets, which are made of ice, or the arrival on Earth of a trail of organic dust left in the path of passing comets. Some researchers, among them the late Carl Sagan at Cornell University and Christopher Chyba at the University of Arizona, have postulated the synthesis of prebiotic molecules in the Earth atmosphere driven by the heat and shock waves caused by impacting bodies (Chyba and Sagan 1992).

Recently, it was suggested that "giant" micrometeorites (50- to 500μm in size, still much smaller than meteorites), which reach the Earth's surface in huge amounts (about 20,000 tons a year), are the dominant current source of extraterrestrial organic material (Maurette 1998). These objects, collected on the Greenland and Antarctic ice fields, were found to contain complex organic compounds, such as amino acids and polycyclic aromatic hydrocarbons (PAHs). Based on the ratios of various chemical isotopes in the micrometeorites, it is estimated that they correspond to large interplanetary dust particles that survived impact with the Earth's atmosphere. These particles were originally ejected by comets approaching the sun (Maurette 1998:404). Michel Maurette and others have suggested an "early-carbonaceous-micrometeorite" scenario, according to which early micrometeorites, similar to those found today, functioned not only as transporters of organic material but also as "microscopic chemical reactors." These reactors supposedly generated complex organic molecules, such as amino acids, on the primordial Earth (Maurette 1998:407–408). The micrometeorites are made of complex aggregates of tiny grains embedded in carbonaceous material. These grains contain clays, oxides, and sulfides of metals, which are known to function as catalysts in various chemical reactions. It is suggested that, upon contact with water, the catalyzed hydrolysis of PAHs could have generated amino acids within the micrometeorites. A similar model was proposed earlier by T. E. Bunch and S. Chang (see Maurette 1998:407) to explain the origin of many organic compounds found in the Murchison meteorite. Furthermore, all micrometeorites are surrounded by a thin shell of magnetite (iron oxide), which could have prevented the dilution of organic materials in the surrounding aqueous solution (Maurette 1998:392, 407).

Interestingly, according to some researchers, the notion of accreting bodies impacting on the Earth after its formation supports the original Oparin-Haldane claim of a reducing atmosphere in which organic synthesis could occur. They argue that planetesimal debris bearing iron metal and carbon accreted to and mixed with the Earth's surface, creating reducing conditions in the mantle and consequently volcanic gases of reduced nature (Gaffey 1997:199, 201). Other mechanisms responsible for the presence of hydrogen and methane, which might lead to a mildly reducing early atmosphere, have also been suggested. One hypothesis envisions the production of hydrogen upon irradiation of solutions of iron salts (containing iron ions) in water, with ultraviolet light as a source of energy. The electrons released from the iron ions in this photochemical process can be donated to hydrogen ions in the water to produce reducing hydrogen gas (Borowska and Mauzerall 1987; De Duve 1991:126–127). Hydrogen could then reduce carbon dioxide and nitrogen gases into methane and ammonia, out of which organic compounds could be produced.

THE HYDROTHERMAL-SYSTEMS HYPOTHESIS

An important process claimed to be of relevance to the possible synthesis of organic compounds on the prebiotic Earth is submarine volcanism (Holm 1992; Corliss 1990). Fluids released from active hydrothermal vents on the seafloor were found to contain mostly carbon dioxide but also substantial amounts of methane (Sakai et al. 1990; Holm 1996). It is contended that the mixture of sulfur minerals and hydrogen sulfide at these sites under high temperatures and pressures produces free hydrogen and free energy capable of reducing carbon dioxide to methane (Wächtershäuser 1992). It is postulated that in the past the Earth's mantle was more reducing and the ratio of methane to carbon dioxide in hydrothermal-vent fluids could have been much higher (Kasting 1997b:1215). Even the presence of a few parts per million of methane would have allowed the synthesis of hydrogen cyanide, which is essential for the prebiotic synthesis of amino acids and nucleotide bases (Kasting and Brown 1996:219). A bolder hypothesis, growing in popularity, is that not only organic molecules but even the first living systems emerged not at the Earth's surface but in the ocean depth (Baross and Hoffman 1985; Holm 1992). Under these conditions, a non-reducing primordial atmosphere would have been no obstacle to emerging life. One of the major proponents of this hypothesis, John B. Corliss of NASA's Goddard Space Flight Center, first detected submarine hydrothermal vents at the Galápagos oceanic ridge in 1979 (Corliss et al. 1979). Corliss and his colleagues argued that the environments of hot springs along ocean-ridge spreading centers could have provided emerging life with both energy and nutrients, as well as with shelter against most extraterrestrial impacts. Several current origin-of-life models postulate prebiotic and early biotic processes taking place on the surface of various minerals under elevated temperatures and atmospheric pressures. These models are claimed to be compatible with the hydrothermal-systems scenario (Pace 1991; Cairns-Smith et al. 1992; Huber and Wächtershäuser 1998).

The deep-sea scenario for the origin of life was suggested after it was discovered by Corliss and others that extant organisms flourish in hydrothermal systems, surviving in total darkness with no sunlight, using instead heat and chemical energy, mainly sulfur compounds emitted by the vents. Support for the new origin-of-life hypothesis came from another source. The evolutionist Carl R. Woese of the University of Illinois at Urbana-Champaign discovered in the 1970s that a group of living microorganisms is the most ancient on the evolutionary scale. These microorganisms prefer extreme conditions—hot environments up to 120°C, anaerobic (oxygen-less) conditions, and high atmospheric pressure. Woese's revolutionary discovery was that these unusual microorganisms—including thermophiles, hyperthermophiles, and halophiles (salt-lovers)—which were formerly classified as bacteria, belong to a third domain or branch of life, which Woese named the Archaea. Though resembling bacterial cells in their lack of a nucleus, archaeal cells have distinct characteristics of their own that can be identified at the molecular level. Thus, in distinction to the traditional division into prokaryotes (cells without a nucleus, such as bacteria) and eukaryotes (cells with a nucleus, including both unicellular algae and protozoa and all multicellular organisms), Woese suggested that life-forms should be divided into three domains: Archaea, Bacteria, and Eukarya (Woese and Fox 1977; Woese 1987). Woese's hypothesis was confirmed in 1996 by the DNA sequencing of a complete archaeal genome (Butt et al. 1996) and its comparison with genomes representing the other two domains (Fleischmann et al. 1995; Williams 1996). Molecular comparisons made it clear that archaeal organisms are located on the oldest part of the tree of life, very close to the root, the last common ancestor out of which all forms of life diverged. Supporters of the hydrothermal-vents scenario assume a direct evolutionary link between the last common ancestor and the first living systems on Earth preceding it. Extrapolating backward to these systems, they infer from the fact that the oldest organisms on Earth are hyperthermophiles that life emerged in a hot environment, possibly in volcanic areas on the seafloor.

Certain other factors among the many suggested in favor of the new origin-of-life hypothesis should be mentioned. The first concerns the time frame within which life is thought to have emerged. It is well established that the Earth was heavily bombarded for at least half a billion years after its formation. Recent indications that life might have originated during this time lend support to the notion that emerging life was somehow protected from impacts. It is conceivable that the deep-ocean environment could have provided such protection. The second supporting factor concerns the physical and chemical dynamics of the vents and of the underlying geological systems near plate boundaries. These sites feature gradients of temperature, pH, and concentrations of various molecules. The thermal and chemical gradients and the flow of compounds among different environments in the vicinity of the vents make the hydrothermal systems highly suitable candidates for processes of organic synthesis (Holm 1992:8). Thermodynamic non-equilibrium conditions are maintained within hydrothermal systems, a fact that enables organic synthesis in these sites (Shock 1992a:92). Based on a long history of organic-synthesis experiments, it has been suggested by geochemists that the high temperatures and pressures in hydrothermal systems should be conducive to the production of organic molecules (Shock 1992b; Kaschke et al. 1994:44).

The discovery of deep-sea vents and microorganisms thriving in extreme environmental conditions has raised hopes concerning the possibility of life on other planets where extreme conditions prevail. Furthermore, the recent putative finding of microfossils on a Mars meteorite, though highly controversial, nevertheless suggests to some astrobiologists that the search for the root of life may lead beyond the Earth. They speculate that the original archaeal organisms might have arrived from space on a meteorite (McKay 1997: 285).

While proponents of the submarine scenario reject the traditional Oparin-Haldane primordial-soup theory, many origin-of-life researchers nevertheless insist on the validity of the older hypothesis. They doubt whether prebiotic organic synthesis could have occurred under high temperatures. Stanley Miller and Jeffrey L. Bada, both at the University of California at San Diego, claim that organic compounds could not have survived the temperatures inside the vents, which might reach 350°C (Miller and Bada 1988; Bada et al. 1994). This view is shared by Antonio Lazcano, an origin-of-life scientist at the National University of Mexico, who notes that amino acids and nucleic acids are unstable at high temperatures and that "modern hyperthermophiles have evolved sophisticated biochemical tricks to adapt to high temperatures" (Balter 1998:31). Supporters of the submarine scenario point out that this criticism is based on a misconception of the nature of hydrothermal systems. Hydrothermal water, they contend, "may have any temperature between 2 and 1200°C" (Holm 1992:13). However, Miller and his colleagues retort that even lower hydrothermal temperatures are too high for the emergence of life as we know it. Having conducted experiments at 100°C that proved RNA nucleotide bases to be unstable, Miller concludes that these molecules could not have accumulated in sufficient amounts under hydrothermal conditions to ensure an origin of life based on nucleic acids (Levy and Miller 1998).

Günter Wächtershäuser, whose theory of the emergence of life on the surface of the mineral pyrite under elevated temperatures and pressures is compatible with the hydrothermal-vents scenario, points out that in surface reactions, unlike reactions in solutions, many thermal degradations are prohibited. Under these conditions a hot origin of life is highly plausible (Wächtershäuser 1992:103).

The idea that life arose in a hot, sulfur-rich environment is also promoted by Christian de Duve, whose theory will be discussed later. De Duve confronts the challenge of the sensitivity of some organic compounds to hydrolysis in hot, acidic conditions by suggesting that "some sort of protection, for example, by a binding catalyst, would have been mandatory" (De Duve 1991:160). Laboratory experiments of organic synthesis under hydrothermal conditions and thermodynamic calculations of the chemical energy that might have been available in the vicinity of hydrothermal systems for organic synthesis indicate that these locations were suitable for the production of organic compounds on the primordial Earth (Simoneit 1995). Moreover, it is also contended that hydrothermal systems were an ideal location for the emergence of the first autotrophic metabolic systems (Shock et al. 1995).

Adding another twist to the story some scientists believe that the first living systems originated in cold-to-moderate temperatures and that hyperthemophilia was a characteristic acquired later as an evolutionary adaptation to changed environmental conditions. According to this scenario, life emerged very early during the bombardment era, but since it was close to the surface it was largely wiped out by major impacts (Zahnle and Sleep 1996). Only those microorganisms that became adapted to hot subsurface environments survived. It is these survivors, not necessarily the first systems to emerge, that we find to be associated with the root of the evolutionary tree (Gogarten-Boekels et al. 1995). Unlike Woese and other researchers, the supporters of this scenario thus question the assumption that both the origin of the first living systems and the evolution of the common ancestor from these systems took place in a hot environment. Another recent report claims that the last common ancestor was not a hyperthermophile. The analysis of genes coding for two ribosomal RNA molecules in forty extant organisms and the construction of a molecular evolutionary tree on the basis of these data revealed that the ancestral genes of all forty organisms could not have survived high temperatures. The nucleotide bases making up these postulated ancient genes were more suitable for moderate temperature. The authors conclude that, in all probability, hyperthermophilic organisms evolved from ancestors living in moderate conditions via adaptation to high temperatures (Galtier et al. 1999). It should be emphasized that this conclusion does not refer to the first living systems but to the common ancestor of extant life on Earth. This hypothetical organism—

at the root of the evolutionary tree—must have already possessed a complex biochemical machinery and is believed to have developed from the first living systems via a complex evolutionary history. A group of leading origin-of-life researchers recently cautioned against the tendency to confuse "origins" and "common ancestor," claiming that "the nature of the last common ancestor, whether or not it was a thermophile, may not provide evidence about 'the cradle of life'" (Arrhenius et al. 1999).

Patrick Forterre, a former supporter of Woese's thermophilic origin-of-life theory, and his colleagues wish to test the hypothesis raised by Woese and others of a direct link between present-day hyperthermophiles and a hot origin of life. They suggest identifying "key molecular features required for life at high temperatures and ask the question: are these primitive traits or adaptive features?" (Forterre et al. 1995:236). A promising way to pursue this research is by comparative sequencing of the genomes of related microorganisms that occupy hot and moderate environmental niches. By identifying and comparing, for instance, the gene that codes for an enzyme that stabilizes DNA against high temperatures in different Archaea and Bacteria microorganisms, researchers hope to trace the evolutionary history of these systems and decide whether the direction of evolution was "from cold to hot or from hot to cold" (Balter 1998:31).

New molecular data based on the sequencing of the entire genomes of more than a dozen microorganisms belonging to the Archaea, Bacteria, and Eukarya appear to be in conflict with the picture of the tree of life originally drawn by Carl Woese. Instead of the clear-cut divisions thought to exist among the three domains of life, the interrelationships that have been discovered among the three groups form a much more complex map. The intricate connections among the various microorganisms tend to obscure the root of the tree, making it hard to pinpoint the universal ancestor common to all living organisms (Doolittle 1998; Pennisi 1998). Woese is now suggesting a new theory according to which the universal ancestor was not a discrete entity, but rather a diverse community of cells. Out of these cells, which evolved by exchanging genes among themselves and improving their ability to synthesize larger and more complex proteins, the three domains of life emerged independently (Woese 1998). The extensive empirical and theoretical work now focused on these issues will no doubt shed light on the conditions and mechanisms involved in the origin of life and the stages leading to the population of cells that constituted the last common ancestor.

Looking at all the hypotheses about the possible sources of organic compounds on the early Earth, it is obvious that, contrary to some claims, the origin-of-life field is not facing a crisis but is rather witnessing a proliferation of new ideas about mechanisms and scenarios that could have been compatible with the currently accepted picture of the prebiotic environment. As indicated by Leslie Orgel, none of the currently entertained ideas on the origin of organic building blocks is compelling, but none can be ruled out. Perhaps, he suggests, several sources of organic molecules collaborated to make possible the origin of life (Orgel 1998:492). The possibility of such a "synergetic effect" between the early-micrometeorite and the hydrothermal-vents scenarios in the production of complex organic compounds on the early Earth was, in fact, suggested recently (Maurette 1998:409–410). Since a huge number of micrometeorites were deposited on the early Earth, including areas where there were hydrothermal vents, these micrometeorites which could have functioned as chemical reactors, might have helped to overcome some of the problems associated with the vents scenario, for instance, the dilution of organic compounds out of the source cores (Maurette 1998:409). Other collaborative ideas between separate, sometimes even conflicting, lines of research in the origin-of-life field will be discussed later.

THE "TIME WINDOW" FOR THE ORIGIN OF LIFE

Intense research has been devoted to the question of when life first appeared on Earth. It is a well established fact that the Earth was formed 4.6 billion years ago. The bombardment from outer space lasted for a long time, probably until 3.8 billion years ago. Because of the enormous impact, the collisions could have heated the oceans enough to boil their surfaces, or even to vaporize them entirely (Sleep et al. 1989). Unless life emerged on the ocean floor and was spared at least some of the impacts, chances are that any incipient organic forms would have been wiped out. Thus the appearance and persistence of life would be delayed at least until the end of the "late heavy bombardment" period. This seems to constitute the lower limit (depending on one's perspective; some people speak of an "upper limit") of what researchers refer to as the "time window"—the time during which processes of prebiotic chemical evolution led to the appearance of the first primitive living system. Judged by the data at hand, the time window was, geologically speaking, extremely short. The lower boundary of the window, as just mentioned, was 4 or maybe even 3.8 billion years ago. The upper boundary is determined by the evidence for the earliest existence of life on Earth, and this date is being pushed back with new discoveries of signs of life in ancient rocks.

Fossils of bacteria cells found in ancient rocks in Africa, aged 2.7 billion years, were considered the oldest evidence available. The estimate thus was that the time window was quite long and lasted for about one billion years. In fact, before evidence of the repeated bombardment of Earth during its first half-billion years was obtained, the accepted estimate was that the emergence of life took a very long time—about two billion years. However, about twenty years ago, J. William Schopf and his colleagues at the University of California, as well as other researchers, have found fossils aged about 3.5 billion years old in sedimentary rocks in Australia (the Warawoona group) and a few other sites around the world. These rocks contain the imprints of strings of bacterium-like bodies, resembling modern microorganisms called cyanobacteria, or blue-green algae. These bodies are embedded in laminated structures called stromatolites, which resemble mats produced by living colonies of bacteria. It should be stressed that these fossils are the remains of highly developed cells that were probably capable of photosynthesis (Schopf 1983; 1993). Another line of evidence for the antiquity of life on Earth involves the ratio between carbon's two stable isotopes in the organic material extracted from rocks. Organisms prefer the lighter isotope of carbon (C-12) over the heavier one (C-13). When organisms fix carbon in the process of photosynthesis, this ratio is enriched in the lighter isotope compared to the ratio in inorganic carbon. The ratio of carbon isotopes extracted from rocks 3.5 billion years old indicates a biological carbon source, which might be due to the process of photosynthesis (Schidlowski 1988).

The microfossils dated about 3.5 billion years might thus be the remains of structurally complex organisms. Assuming that their evolution took time, it can be argued that more primitive life appeared even earlier. The problem faced by paleobiologists is that older rocks have undergone metamorphism, which could have destroyed all previous biological evidence. New techniques, however, make it possible to measure the carbon-isotope composition of minute amounts of carbon-containing material within grains of specific minerals in the oldest known rocks on Earth. It is thought that these mineral grains acted as a shield against the destruction of the carbon-containing material during the process of metamorphism, enabling the present determination of the isotopic ratio of carbon in the rocks. In samples from the oldest sedimentary rocks, aged 3.8 to 3.85 billion years, in the Isua supracrustal belt in West Greenland, the carbon ratio was found to be characteristic of biologically derived material and similar to the ratio obtained from samples in the 2.7-to-3.5-billion-year-old rocks (Mojzsis et al. 1996). Some researchers are still cautious about these results. Similar carbon-isotope ratios could be due to non-biological causes, and the ion-microprobe measurement technique is new and requires more documentation. Nevertheless, it is accepted by most scientists as quite likely that life existed on the Earth by 3.85 billion years ago (Hayes 1996).

The conclusion to be drawn from these results as to the time window is very clear. Life on Earth emerged extremely fast: at most during several hundred million years, or probably during a mere few million years—a blink of an eye in geological terms. It might be argued that life appeared on Earth as soon as the environmental conditions allowed it. Some researchers even claim that, based on the latest findings, there was actually no time between the end of the last bombardment on Earth and the first appearance of life. Since the evidence drawn from the West Greenland rocks overlaps the critical period of bombardment, could it be, they ask, that the organic material in these rocks attests to the fact that not all existing life was wiped out during the collisions? (Mojzsis et al. 1996; Hayes 1996).

I pointed out before that creationists, who believe that life could not have emerged from matter by natural means, regard the extreme shortening of the time window as supporting evidence for their case. (Notice the paradox that the findings of scientific research are seen fit, under the circumstances, to serve as evidence against science.) The underlying philosophical assumption of origin-of-life research, on the other hand, is that life could not have appeared on Earth except by following natural processes of chemical evolution. Guided by this assumption, researchers have to incorporate the new time frame into their theories. Indeed, several of the scenarios suggested today in the field, both in the genetic and the metabolic tradition, describe the early stages of the development of a primitive living system as very rapid (De Duve 1991:216). Antonio Lazcano and Stanley Miller estimate the time for life to arise and evolve to cyanobacteria, "in spite of the many uncertainties involved," to be no longer than "10 million years"! (Lazcano and Miller 1994:546).

GRAHAM CAIRNS-SMITH: THE THEORY OF THE "MINERAL GENES"

Among the scenarios that question and in fact reject, the primordial-soup notion, offering to replace the Oparin-Haldane hypothesis as the framework for origin-of-life research, the theory formulated by the chemist Graham Cairns-Smith of the University of Glasgow is of particular interest (Cairns-Smith 1985). Following the changed conception of the prebiotic atmosphere, it is evident, according to Cairns-Smith, that organic compounds could not have accumulated on the early Earth as required by the conventional model of chemical evolution (Cairns-Smith 1985:6, 42–43). Cairns-Smith does not share the optimism of those who postulate a reducing atmosphere resulting from the release of free hydrogen from water in the presence of iron salts, under the effect of ultraviolet radiation. He points out that ultraviolet radiation is also responsible for other processes that tend to produce an oxidizing effect. Thus he supports the notion of the emergence of life as far away as possible from the sun's radiation, for instance, in hydrothermal systems at the bottom of the sea. Such an environment could provide reducing conditions in the form of different minerals within and beneath the Earth crust (Cairns-Smith et al. 1992:162).

Many origin-of-life scenarios ascribe diverse roles to minerals in the prebiotic environment. However, Cairns-Smith's is among the most far-reaching and radical. According to him, it was mineral chemistry and not the organic chemistry of carbon compounds that brought about the emergence of life. Organic chemistry, Cairns-Smith believes, is much too complex and not at all suitable for the initial stages of the process. Though a few organic substances—for instance, certain simple amino acids—can form relatively easily under prebiotic conditions, other biochemical building blocks, such as nucleotides and lipids, require for their synthesis a "real factory" (Cairns-Smith 1985:48). The synthesis of these substances involves a series of reactions, each reaction following the previous one in the utmost accuracy. Cairns-Smith denies the possibility that such a chain could be a product of chance on the ancient Earth. On the basis of several calculations, he contends that "there was not enough time, and there was not enough world" for that (Cairns-Smith 1985:47). He concludes that a natural origin of life, in distinction to a miraculous origin, could not have started with proteins, nucleic acids, sugars, and the like. If life indeed started without the help of miracles, the first organisms must have been made of materials that were easily available. It is Cairns-Smith's suggestion that the first "organisms" were made of mineral crystals, such as different clays, which are among the most ubiquitous materials on Earth. The chemistry of emerging life was inorganic rather than organic, and hence much simpler. Cairns-Smith believes that this is a manifestation of a general principle of innovation, technological or evolutionary: such a process starts with what is easiest and most accessible and later moves to the more complex and efficient (Cairns-Smith 1985:115–116; Cairns-Smith et al. 1992:163).

In the model suggested by Cairns-Smith, the first inorganic organisms served as a scaffolding on which life as we know it, built of organic compounds, could later develop (Cairns-Smith 1985:58–61). Cairns-Smith draws an important distinction between "living entities" and "organisms." Living entities are the products of evolution, while organisms are prerequisites for evolution. An organism is a system capable of evolving through natural selection (Cairns-Smith 1985:3). Minerals, he claims, fulfill this definition. It is possible to describe clay minerals, he says, as "crystal genes." "A crystal gene must be able to hold substantial amounts of information, and replicate that information rather accurately through processes of crystal growth and crystal break-up. And . . . this information must have some effect . . . or effects that help the genes holding that information to survive better, or replicate faster, or be spread around more widely" (Cairns-Smith 1985:74–75). Information pertaining to crystal genes is embodied in the specific distribution of electrical charges on their surface due to irregularities in their structure. These genes, he contends, were able to replicate, direct the synthesis of a new layer on the mineral surface, and in that way transmit their irregularities—the information—to the "next generation." Due to mistakes—"mutations"—during this process, different structures characterized by different properties were formed. Since the distribution of charges on the mineral's surface could determine, among other things, the efficiency of its copying, natural selection among the crystal genes could ensue. In such a manner, by the selection of the "fitter" crystal genes, evolution of more stable and complex structures better fitted to their environment was able to occur (Cairns-Smith 1985:98–106).

The crucial point Cairns-Smith makes is that during this evolutionary process, the genes learned how to synthesize organic compounds on their surfaces. These compounds provided the genes with different environmental advantages. In this manner, different organic molecules were gradually built, until eventually the organic genes—nucleic acids—were much superior in stability and efficiency to the inorganic genes, and a "genetic takeover" took place. Thus, at a certain historical point, there was no longer need for the mineral scaffolding, which was discarded, leaving no trace of its previous existence behind (Cairns-Smith 1985:107–113).

As yet, there are no experimental indications to support Cairns-Smith's model. Leslie Orgel finds its difficult to believe that structural irregularities in clay complicated enough to produce organic polymers similar to RNA could be responsible for accurate self-replication (Orgel 1994:61). Most critics of Cairns-Smith's scenario focus on the fact that his presumed ancient scaffolding did not leave any mark that can be detected in present organisms. Indeed, the claim that life did not start at all with organic chemistry challenges a basic "continuity conception" guiding research into the origin and evolution of life. According to this conception, every stage in evolution is connected both to the stages preceding it in the early development of life and to those that followed in the development of present-day organisms (Morowitz 1992:27). Many researchers believe that several of the most basic characteristics of the biochemistry of the living cell known to us today emerged very early in the development of life. When a specific structure was formed, it was highly beneficial to preserve it and very risky to change it. In this context, the term "molecular fossil" refers to any molecule or function that was "frozen in time," since any change would have entailed an enormous number of simultaneous changes that could have brought about a system's demise (Weiner and Maizels 1991:53). For this reason many scientists believe that the primordial organization of life could not have been totally different from the present nucleic acid–protein system, and thus they reject Cairns-Smith's basic assumption.

In response, Cairns-Smith claims that challenging the continuity conception by postulating an inorganic scaffolding is the only way to explain the emergence of life on the basis of natural processes. The conventional organic scenario, he says, has to assume processes of such improbability that recourse to highly singular events or to miracles becomes necessary (Cairns-Smith 1985:6–8; Cairns-Smith et al. 1992:161). It is nonetheless evident that despite the controversy, Cairns-Smith and the "continuity people" share a common philosophical view in opposition to creationists: the idea that specific mechanisms were responsible for the natural emergence of life from matter, and that these mechanisms can be discovered by origin-of-life research.

The recent application of Cairns-Smith's theoretical claims to the hydrothermal scenario may offer a promising boost to origin-of-life research (Russell et al. 1988; Cairns-Smith et al. 1992). Looking in the hydrothermal environment for mineral assemblages from which primitive genes and catalysts as well as membranes of the first inorganic organisms might have been made could provide an empirical basis for Cairns-Smith's theory. It could also extend this theory beyond the scope of a genetic theory by pointing out that mineral genes, catalysts, and membranes together constitute the minimal requirement for emerging life. Cairns-Smith's idea of a genetic takeover could then be extended to include the change at a later evolutionary stage from mineral catalysts to peptides and from mineral membranes to lipids (Cairns-Smith et al. 1992; Holm et al. 1992:188).

CONSERVATIVE ROLES FOR MINERALS

In distinction to Cairns-Smith's radical theory, many researchers have suggested more conservative mineral theories, not disputing the basic assumption that life began with organic chemistry but still considering minerals as playing an important role in the emergence process. Bernal pointed out as early as 1951 that the low concentrations of organic compounds in the dilute primordial soup pose a problem for the chemical reactions involved in the emergence of life. He suggested that different minerals, especially clay minerals, that were abundant on the sea bottom could have overcome this problem by adsorbing organic molecules on their surfaces and thus concentrating them (Bernal 1951). In addition to adsorption, Bernal and others raised the idea that minerals could be instrumental in the synthesis of organic polymers. Indeed, one of the most difficult problems associated with the soup scenario involves the synthesis of polymers like oligonucleotides and peptides (short chains of nucleic acids and proteins) in aqueous solution: linking the building blocks, amino acids or nucleotides, into a polymer requires the extraction of a water molecule from two adjacent units, a difficult task in the presence of water. The favored reaction under such conditions is hydrolysis, the separation of units through the addition of water, which obviously conflicts with polymerization. By adsorbing the building blocks on the mineral surface, the processes of dehydration and polymerization are favored over hydrolysis (Ferris 1987; Von Kiedrowski 1996).

Measuring the rate of several reactions, among them various polymerizations, on the surface of minerals clearly shows that often the minerals act as inorganic catalysts. The fixation of different molecules, about to interact, on the surface facilitates bond formation and hence speeds up the reaction (Ferris et al. 1990; Ferris and Ertem 1992; Von Kiedrowski 1996). Experiments conducted recently by James Ferris and his colleagues at the Rensselaer Polytechnic Institute, as well as by other researchers, indicate, moreover, that various minerals tend to bind the reaction products, the formed polymers, and this binding, mainly by electrostatic interactions, increases with the length of the chain. The bound polymers are better protected from disintegrating back to their constituents through hydrolysis. Using the common montmorillonite clay, Ferris and Orgel succeeded in synthesizing an oligonucleotide up to fifty units long. Polymers containing more than fifty units of amino acids were synthesized on the surface of the minerals illite and hydroxylapatite, though in this reaction the mineral did not act as a catalyst (Ferris et al. 1996).

A recent report by Ferris and Gözen Ertem indicated that oligonucleotides formed by the catalysis of a montmorillonite clay can serve as a template and direct the synthesis of another polymer with complementary bases—the first step in the process of replication (Ertem and Ferris 1996). Not only is the clay mineral acting as a catalyst in the polymerization of oligonucleotides on its surface; it might also act as a catalyst in the template-directed synthesis of the complementary oligonucleotides. In this reaction, catalysis by minerals may result in polymers with defined structures, in comparison with more heterogeneous oligonucleotides resulting from a non-catalyzed template-directed synthesis.

Several researchers have attributed to minerals the ability to protect organic compounds from ultraviolet radiation. Others describe adsorption on mineral surfaces as equivalent to the formation of a niche protected from the environment: the mineral is supposed to function as a membrane surrounding a sort of primordial cell (Wächtershäuser 1992:104–108). Several of the most difficult problems in the soup scenario involve the presence of many "wrong" molecules in the soup, which might inhibit the "right" reactions. This crucial question will be discussed in the following chapters. However, it should be pointed out that some models regard the ability of minerals to favor the adsorption of some chemical compounds over others as a means to overcome that difficulty (Bonner 1991; Orgel 1994:57–60).

In summary, most scenarios that allot to minerals important roles in the emergence of life are not as far-reaching as Cairns-Smith's. They do not deny the crucial role of organic chemistry, and most of them do not give up the notion of the primordial soup, despite the many problems it brings with it. These scenarios still rely on the aqueous solution of the soup as a source for the organic building blocks that adsorb and interact on the mineral surface (Von Kiedrowski 1996).

THE "DIRECTED-PANSPERMIA" THEORY

The realization that the time window during which life could have formed on Earth was much shorter than previously assumed inspired suggestions of rapid scenarios. It also gave rise to a renewed interest in the notion of panspermia, the transfer of life from outer space to Earth. Uncertainty regarding the composition of the primordial atmosphere also encourages suggestions that at least certain stages of the process through which life emerged occurred in space (Goldanskii and Kuzmin 1989; Gribbin 1993). The ideas of Francis Crick about the emergence of life are of particular interest here. Crick dealt with the difficult empirical question of how complex organic molecules could have been synthesized on the primitive Earth, assuming that the atmosphere was not as reducing as previously thought, by suggesting that "planets elsewhere in the universe may have had more reducing atmospheres, and thus have on them a more favorable prebiotic soup" (Crick 1981:79). Crick's speculations led him much farther in his attempt to solve the difficult problem of the origin of life on Earth. First in 1973, with the origin-of-life chemist Leslie Orgel (Crick and Orgel 1973), then in 1981, in *Life Itself*, and later in his autobiography, *What Mad Pursuit*,

Crick presented his "directed-panspermia" theory on the emergence of life, according to which "life on Earth originated from microorganisms sent here, on an unmanned spaceship, by a higher civilization elsewhere" (Crick 1990:148).

On Crick's testimony, he and Orgel were originally led to this wild idea by two facts: first, the uniformity of the genetic code and hence the need to assume the evolution of life through a "narrow bottleneck." Second, because the age of the universe is more than twice the age of the Earth, it is conceivable that life has evolved twice from simple beginnings to highly complex intelligence—first on another planet and then on Earth (Crick 1990). According to comments by colleagues and reporters, Orgel and Crick do not take their theory very seriously; in fact, Orgel has said that it was "sort of a joke" (Horgan 1991:125). Their intention, though, was serious: to call the attention of both the public and scientists to the enormous difficulties involved in the explanation of the emergence of life on Earth (Shapiro 1986:227). As for the directed-panspermia theory itself, it should be pointed out first that since obviously Crick, unlike the panspermists at the end of the nineteenth century, does not believe in the eternity of life in the universe, the transfer of the emergence-of-life arena to another planet does not solve the problem of origins. It simply postpones it, transferring it to another planet. Second, and more important, analysis of the philosophical assumptions underlying the directed-panspermia theory shows that it is not by chance that Crick's scenario holds onto "a trace of magic," to use Cairns-Smith's phrasing. Discussing the assumption, until quite recently common among scientists, that the emergence of life was a highly improbable event, in fact a singularity, Cairns-Smith comments: "There is a temptation, in any case, to suppose that if the origin of life was not actually supernatural it was at least some very extraordinary event, an event of low probability, a statistical leap across a great divide. That way a trace of magic can be held onto" (Cairns-Smith 1985:2).

Crick raises the challenging question, How likely is it that a system that could have later evolved by natural selection arose spontaneously? (Crick 1981:80). He assumes that this process involved a sequence of rare events for which experimental support is very difficult to attain, and thus "we cannot decide whether the origin of life on Earth was an extremely unlikely event or almost a certainty, or any possibility in between these two extremes" (Crick 1981:88). And yet Crick's inclination is toward the first "rare-event" possibility. Several of his assertions about the origin of life indeed express the feeling that we are dealing here with an extraordinary event: even in a hypothetical more favorable environment on another planet, Crick still regards the emergence of life as a "happy accident" that only the passage of millions of years helped to bring about (Crick 1981:39). An honest man, he claims, armed with all the knowledge available to us now, could only state that in some sense the origin of life appears at the moment to be "almost a miracle, so many are the conditions which would have had to be satisfied to get it going" (Crick 1981:88).

The scenario postulated by the directed-panspermia theory was inspired by the difficulties facing origin-of-life research. Yet relegating the question to another planet does not solve the problem, which, I claim cannot be solved as long as the notion of a "happy accident" is entertained. First there is the empirical question, What is the probability of the occurrence of physicochemical conditions favorable for the emergence of life on Earth or elsewhere? Beyond this there is the further question, Can matter under any conditions, even those most conducive to life, organize itself into a biological system? This question, which involves our most basic conceptions of life and matter and their interrelation, is sometimes given the "happy accident" answer. The fact that life exists on Earth, it is claimed, is proof that life emerged from matter. However, the chances for such an event are extremely small, or "virtually zero," to recall the remark of Jacques Monod (Monod 1974:136). Though obviously neither Crick nor Monod denies that life arose by natural processes, the idea that it arose as the result of a series of very rare events leads to a scientific dead end. Crick's "almost a miracle" is often quoted by creationists in their attempt to portray the entire origin-of-life research project as bankrupt (Bradley and Thaxton 1994:191; Johnson 1993:110–111). Arguing that life could not have emerged through a series of chance events, they contend that origin-of-life theories have not come up with any other viable mechanism (Bradley and Thaxton 1994:191–197).

Toward the end of the book, I will discuss another panspermia theory, much more far-reaching in its claims than Crick's, suggested by the British astrophysicist Fred Hoyle and his colleague Chandra Wickramasinghe. Hoyle believes that the amount of information required for a biologically organized system could not have emerged through natural processes, even under favorable conditions in outer space. The conclusions he and his colleague reach are thus religious in nature. In a completely different vein, we have already noted that, following the controversial discovery of putative microfossils in a Martian meteorite, the possibility that microorganisms were transferred to Earth by meteorites ejected from Mars has been discussed (McKay 1997). It is further speculated that these microorganisms could have been responsible for the beginning of life on Earth. Unlike the claims made by Hoyle and Wickramasinghe, the philosophical assumption underlying such panspermia ideas is that life emerged on Mars or another planet by natural processes, and that it can undergo processes of self-organization wherever physicochemical conditions are appropriate.

Though the idea of life on Earth being seeded by an extraterrestrial source cannot be ruled out, current origin-of-life research is focused on possible terrestrial mechanisms that could have led to the emergence of life. Recent developments in the investigation of hydrothermal systems and their potential for the origin of life are an example of this fertile approach. Other examples will be discussed in the following chapters.

11

THE RNA WORLD— A CASE FOR RENEWED OPTIMISM?

Francis Crick has suggested, seriously or jokingly, the directed-panspermia theory. Graham Cairn-Smith put forward the theory of mineral genes. Implicit and explicit creationists speak of a new "science" that will solve the origin-of-life problem by putting God into the picture. Though belonging to radically different philosophical categories, all these theoretical attempts challenge the conventional model of the emergence of life on Earth based on the Oparin-Haldane hypothesis. Most researchers in the field, however, prefer to face the difficulties involved in this model not by abandoning planet Earth, not by forsaking organic chemistry, and certainly not by rejecting the scientific worldview. And yet the difficulties are many and varied, and in addition to all those already mentioned, such as the non-reducing nature of the primordial atmosphere, the drastic shortening of the time window, and the dilution of the primordial soup, there is another major difficulty to be tackled. I have already referred to it as the chicken-and-egg problem: since in every known living cell nucleic acids and proteins are completely interdependent as far as synthesis and activity are concerned, the question arises as to how this "vicious circle" first emerged. The postulation that nucleic acids and proteins, being highly complex, arose at the same time is highly improbable. How, then, did the first proteins emerge without nucleic acids, or alternatively, how were nucleic acids first synthesized and how could they replicate without proteins?

It is the second question that interests most researchers in the field, who consider proteins to be an unsuitable candidate for independent emergence. Indeed, in the last decades the genetic conception—according to which the first to arise were nucleic acid–like polymers able to undergo evolution through processes of replication, mutation, and natural selection—became the dominant line of research on the origin of life. Already in the late 1960s, Crick, Orgel, and Woese developed independently the hypothesis that a primitive system consisting exclusively of RNA could possibly have existed on the primordial Earth (Crick 1968; Orgel 1968; Woese 1967:179–195). Assuming that such an RNA system could have performed all the complicated activities now carried out by nucleic acids and proteins, this idea was supposed to resolve the chicken-and-egg problem. The claim was that unlike the present situation, where there is a division of labor between molecules carrying information—the nucleic acids—and molecules functioning as enzymes—the proteins—in the past, RNA could have functioned at the same time both as carrier of information and as an enzyme. On the primordial Earth, then, RNA served as both chicken and egg.

According to this hypothesis, when life emerged, RNA performed two major enzymatic activities. First, it functioned as a replicating enzyme and replicated itself without a protein. Second, at a later stage, RNA started to catalyze the different processes involved in protein synthesis. Gradually, following processes of natural selection, the proteins synthesized in this manner became the efficient enzymes known to us today and could replace the RNA enzymes. Later, the change from RNA to DNA as a carrier of information took place (Gilbert 1986).

This hypothesis relied on the fact that in most modern biological systems RNA performs several key functions both in DNA replication processes and in protein synthesis. It is RNA and not DNA that was suggested as the original actor in the prebiotic drama because, first, RNA serves as the genetic material in many kinds of viruses. Second, it is much easier to synthesize the building blocks of RNA, the ribose-containing nucleotides, than the DNA nucleotides. In the cell, the building blocks of DNA are indeed synthesized from those of RNA and, thus it is possible to imagine how the more stable DNA molecule evolved later and replaced RNA as the information molecule (Joyce 1989:217–218). The claim that in the past RNA could function as an enzyme was based also on the spatial structure characteristic of this molecule. Unlike DNA, which usually forms a stable double-stranded structure, the double helix, RNA consists in most cases of a single polynucleotide chain, folding up into three-dimensional structures that could manifest active sites (Cech and Uhlenbeck 1994; Michel and Westhof 1996). The significance of this fact will be better understood once I describe a few basic features of enzymatic activity—in particular, the connection between the spatial structure of protein chains and the ability of enzymes to catalyze specific biochemical reactions.

Like any other catalyst, an enzyme, which is a biological catalyst, increases the rate of a chemical reaction without itself undergoing any change. It should be pointed out that most biochemical processes could not occur under the prevailing conditions in the cell without the participation of enzymes, which speed up the rate of biochemical reactions at least ten million–fold. As already mentioned in chapter 8, an essential first step in enzymatic catalysis is the association between an enzyme and its substrate, a compound about to undergo some chemical change. It is this association that facilitates the chemical process, at the end of which reaction products are released from the enzyme. The binding between the enzyme and its substrate takes place in the enzyme's binding site. This site comprises several amino acids that might be located far from each other on the protein chain. Through the folding-up of the enzyme protein chain, these amino acids come into close contact and can form the three-dimensional structure of the binding site. The association of substrate and binding site is highly specific, and might be compared to the fit between a lock and its key. All the enzymes known to us until recently are proteins that manifest very elaborate three-dimensional structures, which guarantee a specific and efficient combination of each enzyme with its substrate.

THE DISCOVERY OF RIBOZYMES—RNA ENZYMES

The ability of RNA molecules to fold up into intricate three-dimensional structures suggested the possibility of catalytically active sites on these molecules. Yet in light of the long-standing identification between enzymes and proteins, the idea that a nucleic acid could have functioned in the past as an enzyme was considered far-fetched. Surprisingly, it was discovered recently that RNA enzymes are not just hypothetical entities in an origin-of-life theory, but actually exist today. Two American researchers, Thomas Cech of the University of Colorado and Sidney Altman of Yale University, discovered independently, in the early 1980s, enzymes made of RNA in present-day organisms (Kruger et al. 1982; Guerrier-Takada et al. 1983). Cech and Altman, who were awarded the Nobel prize in 1989, called these enzymes "ribozymes," a combination of the terms RNA and enzyme. The ribozymes discovered by Cech and Altman catalyze the cutting and joining of segments of RNA. Though these functions are less sophisticated than those associated with RNA replication and protein synthesis, the discovery of ribozymes nevertheless lent significant strength to the hypothesis of the role played by RNA in the emergence of life. In 1986, Walter Gilbert, a biologist at Harvard University, coined the term "RNA world" to refer to the chemical world prevailing on the ancient Earth, in which RNA molecules were responsible for their own replication, and later also for the synthesis of proteins and DNA (Gilbert 1986).

Other researchers have subsequently found evidence of different enzymatic activities catalyzed by RNA. A major finding concerns the ribosomes—subcellular structures that function as the protein-synthesizing machinery of the cell (Note the distinction between "ribozymes" and "ribosomes.") Ribosomes are composed of both protein and ribosomal RNA, and the dominant assumption until recently was that the protein components in a ribosome are responsible for protein synthesis. This assumption was called into question when H. F. Noller, Jr., of the University of California at Santa Cruz discovered that ribosomal RNA seems to catalyze the formation of peptide bonds between amino acids in the process of protein synthesis (Noller et al. 1992). There is also evidence that an RNA component of the ribosome isolated from another microorganism is active in yet another crucial step in protein synthesis, involving the binding between an amino acid and transfer RNA, the RNA responsible for the transferring of amino acids to the ribosomes (Piccirilli et al. 1992).

No evidence has yet been found of the present existence in nature of RNA molecules capable of catalyzing their own replication. Yet the search for such activity is being carried out through sophisticated experiments exploring evolution of RNA molecules in the laboratory. I have already described the first attempts at evolution in a test tube of the Qß virus RNA that were performed by Sol Spiegelman in the 1960s. Through repeated cycles of replication,

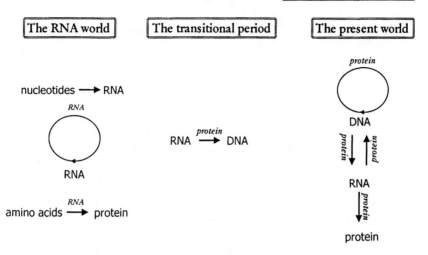

FIGURE 6. *The RNA world and the transition to the present DNA-RNA-Protein world.*

The RNA world. RNA synthesized on the primordial Earth functioned as both a carrier of information and an enzyme. It catalyzed its own replication and also the synthesis of proteins.

The transitional period. Proteins previously synthesized with the help of RNA enzymes catalyzed the transition from RNA to DNA.

The present world. In life as we know it, protein enzymes catalyze the replication of DNA. They also catalyze the transcription of DNA into RNA and the translation of messenger RNA into proteins.

mutation, and selection, Spiegelman, by changing the "environmental conditions" in the test tube, could select specific mutants out of the many produced when the enzyme Qß replicase catalyzed the replication of the viral RNA. Based on the same principles, Darwinian evolution has been carried out in vitro in recent years with the purpose of generating new drugs, industrial chemicals, and also new ribozymes (Joyce 1992; Ellington et al. 1997). This line of investigation, intensively explored by Jack W. Szostak and David P. Bartel of the Massachusetts General Hospital and by Gerald F. Joyce of the Scripps Research Institute as well as by many other researchers, has proved to be very successful, leading to the isolation of new ribozymes with a variety of catalytic functions (Wilson and Szostak 1995; Wright and Joyce 1997).

The application of this technique to the study of the RNA-world hypothesis starts with the preparation of a large pool of random-sequence RNA molecules, from which variants with weak catalytic properties are picked out. These variants are replicated under mutagenic (mutation-generating) conditions, using a protein enzyme and the required building blocks, the four nucleotides. Selection is then carried out to identify new and more powerful variants. Following this procedure of sequential cycles of replication, mutation, and selection, the evolution of ribozymes with new functions or higher catalytic efficiency is achieved (Joyce 1992). The isolation of new RNA enzymes from a random pool of RNA sequences is compatible with the logic behind the concept of the RNA world. According to this logic, the first ribozymes arose from a prebiotic pool of random RNA sequences, including an RNA replicase, that is, an RNA molecule capable of catalyzing its own replication (Bartel and Szostak 1993:1411, 1417). All natural ribozymes discovered so far catalyze only a restricted number of reactions. However, the list of enzymatic activities of RNA molecules isolated in vitro is much longer, including catalytic functions other than those involving the RNA backbone. Supporters of the RNA-world scenario believe that this points to possible prebiotic enzymatic functions of RNA molecules that have disappeared from the biological scene, being displaced in the course of evolution by more efficient protein enzymes (Wilson and Szostak 1995:777).

The RNA-world scenario, which was intended to resolve the chicken-and-egg problem, postulated that RNA preceded the emergence of proteins and that the first self-replicating system to emerge was made of RNA. This scenario was based on highly optimistic assumptions concerning prebiotic chemistry, such as the uncomplicated synthesis of the building blocks of RNA, nucleotides, and of their components and the spontaneous assembly of these building blocks to form RNA sequences. The scenario also suggested that at least one of the sequences had the properties of a ribozyme that was able to self-replicate. Following many recent experiments in prebiotic chemistry, however, there is a growing number of dissenting voices calling this optimistic picture into question (De Duve 1991:128–130; Joyce 1992; Orgel 1994:57–61). Interestingly enough, several of the major supporters of the genetic conception, among them leading researchers in the chemistry of RNA, have emphasized the difficulties involved in the optimistic scenario of the RNA world. They do not question the critical role played by RNA in the emergence of life, but they doubt whether it could have been the first self-replicating system to emerge. In their view, origin-of-life research should focus on the questions, What preceded the RNA world, and what were the plausible chemical mechanisms that led to its emergence? (Joyce and Orgel 1993).

Many researchers note the difficulties that would have been involved in prebiotic synthesis of the RNA sugar component, ribose, and of a few of the nitrogen bases (Shapiro 1988; 1995). Assembling the different components of the nucleotides—sugars, nitrogen bases, and phosphates—under these conditions must have also posed serious problems. In all probability, a great variety of chemical reactions took place in the primordial soup, both "correct" and "incorrect" ones. In many cases the incorrect products could have interfered with and inhibited the desirable reactions (Joyce 1992:220–222). Orgel and Joyce claim that even assuming that the correct building blocks were somehow produced and that they were somehow spontaneously assembled to form an RNA polymer, it is difficult to imagine that among the products was an active, self-replicating ribozyme (Joyce and Orgel 1993:11). They point out that many conditions were required for an active replicating ribozyme to be formed and that these conditions were extremely difficult to satisfy in the prebiotic environment. An active polymer would have to be of a certain minimal length; it would have to contain specific building blocks, arranged in a particular order; the chemical bonds between the components would have to be of the correct kind. Only if some of these demands were met could such a sequence serve as a template for the attachment of complementary units and form three-dimensional structures that would make enzymatic activity possible (Joyce and Orgel 1993:7–11).

Satisfying these stringent demands becomes even harder to imagine, according to Joyce and Orgel, once it is realized that the synthesis of one such correct sequence is not enough. Since RNA was supposed to function as both chicken and egg, unless the RNA molecule could literally copy itself, two identical, or very similar, polymers were required: one was supposed to unfold and act as a template, and the other had to fold up and act as an enzyme, catalyzing the replication of the template. The probability that two such molecules could be found among a random pool of RNA sequences is so slim that, according to the calculation of Orgel and Joyce, the amount of RNA required would be greater than the mass of the Earth! (Joyce and Orgel 1993:11). There seems to be a way out of the dilemma, based on the experiments leading to the evolution of new ribozymes. It seems reasonable to assume that a prebiotic ribozyme functioning as an efficient replicase could have evolved from a less efficient RNA replicase, following cycles of replication, mutation, and selection.

Joyce and Orgel, however, claim that here we encounter another chicken-and-egg problem: "Without evolution it appears unlikely that a self-replicating ribozyme could arise, but without some form of self-replication there is no way to conduct an evolutionary search for the first, primitive self-replicating ribozyme" (Joyce and Orgel 1993:13). Since what was needed was a starting point, some preliminary form of self-replication, this second vicious circle could have been resolved had it been possible to achieve primitive self-replication with no need for an enzyme, either a protein enzyme or one made of RNA. I have already noted the experiments conducted in the 1960s and 1970s by Leslie Orgel and his group in which the possibility of replication without enzymes was explored. Synthesizing oligonucleotides and mixing them with free nucleotides, Orgel did succeed in producing complementary chains of oligonucleotides on the original template. A template made of uracil-bearing nucleotides directed the synthesis of a short strand of adenine-bearing nucleotides, and oligonucleotides containing units of cytosine directed the synthesis of an all-guanine oligonucleotide. Orgel thus succeeded in what he described as the "copying" of the template chain. However, further experiments in which the second stage in the process of replication—the copying of the copy synthesized in the first stage—was attempted did not succeed. To produce a duplicate of the original RNA sequence, a protein enzyme was needed (Orgel 1994:60). If non-enzymatic replication did not work, how then could an original RNA self-replicating system have emerged?

WHAT PRECEDED THE RNA WORLD?

Facing the difficulties involved in the prebiotic synthesis of ribonucleotides and an RNA molecule capable of self-replication, many chemists tend to believe that RNA was not, after all, the first self-replicating molecule to emerge, but may have been preceded by simpler replicating molecules. These simpler systems, following processes of evolution and natural selection, eventually gave rise to the RNA world. A notable example previously mentioned, suggested by Graham Cairns-Smith, is a genetic system made of clay minerals, which was eventually "taken over" by a more sophisticated genetic system made of nucleic acids. Unlike Cairns-Smith's, most efforts to find a simpler genetic system concentrate on alternative organic molecules. The idea is to try to circumvent the problems involved in the synthesis and function of RNA by relying on RNA-like molecules that contain substitutes for or analogues of ribose and some of the nitrogen bases (Orgel 1994:60–61). The Swiss chemist Albert Eschenmoser of the Federal Institute of Technology (ETH) at Zürich has synthesized a molecule slightly different from ribose, rearranged to include five carbons and one oxygen atom rather than the four carbons and one oxygen atom included in the ribose ring. The RNA-like strands containing this component, called pyranosyl-RNA, can pair according to the Watson-Crick complementarity rules characteristic of regular nucleotides. This structure thus fulfills the basic requirement of being able to function as an informational system. Moreover, in preparation for replication, the paired modified strands separate more easily than regular double strands of RNA, which tend to twist around each other (Eschenmoser 1994). In an extensive study carried out by Stanley Miller and his group, a five-membered ring substitute for the six-membered pyrimidine base uracil, called urazole, as well as other five-membered nitrogen bases, was produced and found to be easily linked to ribose to form nucleoside analogues. It is also postulated that since urazol does not absorb ultraviolet light at wavelengths where uracil absorbs strongly, RNA precursors containing urazole might have been less sensitive to ultraviolet damage than uracil-containing RNA (Kolb et al. 1994).

Another candidate for a pre-RNA genetic molecule, called peptide-nucleic acid (PNA), is supposed, among other things, to address a major difficulty—the problem of chirality—involved in the RNA-world scenario. In my discussion of Louis Pasteur's investigations in the nineteenth century, I referred to his discovery of the fact that many organic molecules within the living cell can be arranged in two different spatial structures that manifest opposite optical activity, each rotating the plane of polarized light in an opposite direction. These optical isomers, later designated L and D, structurally relate to each other like an object and its mirror image, or like the right and left hands, and thus the phenomenon of asymmetry is called "chirality" after the Greek word *cheir*, or hand. A yet unsolved puzzle is the fact that both nucleic acids and proteins in all living systems are homochiral: with rare exceptions, all amino acids in proteins are of the L, or left-handed, type, and all sugars in the backbones of DNA and RNA are of the D, or right-handed, type. The questions asked by origin-of-life researchers in this regard are, Why are all amino acids and sugars in all organisms of the same chiral type, and why was one isomer "chosen" over its mirror-image, or "enantiomer"? Was homochirality a precondition for the emergence of life, and did the mechanism responsible for homochirality precede the appearance of living systems on Earth? Or did homochirality arise later as a result of the process of evolution that led to the common ancestor? (Hegstrom and Kondepudi 1990; Bonner 1991; Cohen 1995a). The many hypotheses concerning the origin of the homochirality of proteins and nucleic acids can be roughly divided into physical and biological ones. Physical theories postulate a physical process that occurred on Earth or in outer space prior to the origin of life, assuring a supply of homochiral organic building blocks (Kondepudi and Nelson 1985; Bradley 1994; Greenberg 1995). Biological theories, on the other hand, claim that the chiral character of biomolecules was not a prerequisite but rather a product of life, resulting from some selective advantage of one of the optical isomers randomly formed in the process of life's emergence (Orgel 1973:165–168; Bada 1995).

These questions are of the utmost importance when we deal with the chemistry of the RNA world and the pre-RNA world. It was already pointed out by Louis Pasteur, more than a century ago, that unlike chemical reactions within the cell, synthesis of organic compounds outside the cell always produces a racemic mixture, that is, an equal mixture of the two optical isomers. According to the RNA-world scenario, all the building blocks, including the sugar ribose, were produced by chemical reactions taking place in the primordial soup. It is thus reasonable to assume that the soup contained a mixture of both D and L ribose. However, experiments reported by Joyce and Orgel in which the copying of an oligonucleotide template was attempted seem to show that copying can be achieved only when all the nucleotides are of the right-handed type. Addition of left-handed nucleotides to the reaction mixture inhibited the process (Joyce et al. 1984; Joyce and Orgel 1993:19). It is thus difficult to visualize how, in a racemic prebiotic environment, RNA-world reactions could have taken place.

Peptide-nucleic acid, or PNA, mentioned above as a candidate for an RNA alternative, was supposed, among other things, to tackle this very problem (Piccirilli 1995). The PNA molecule was designed and synthesized by the chemist Peter E. Nielsen of the University of Copenhagen. It is a polymer consisting of a peptide backbone to which nucleic-acid bases are attached. The protein-like backbone, replacing the ribose-phosphate backbone of RNA, is composed of units of glycine, the only amino acid that is achiral—has no asymmetric carbon and hence no optical isomers. Assuming that such an achiral polymer could have been synthesized in a prebiotic environment, it would not have faced the problems involved in the copying of chiral polymers. Sequences of PNA form base pairs in a similar fashion to RNA, as well as double-helix structures (Nielsen 1993; Wittung et al. 1994). Recent experiments indicate that both RNA and DNA segments can serve as templates for PNA, and vice versa. These results point to the possibility of the transfer from a pre-RNA world to an RNA world, or perhaps even to a DNA world (Böhler et al. 1995).

THE ARGUMENT FOR THE "REAL THING"—NATURAL NUCLEOTIDES

Notwithstanding the potential ability of various RNA substitutes to self-replicate, it is nevertheless the claim of some chemists that the first self-replicators were RNA molecules, after all. They are not willing to give up the distinct advantages of nucleic acids as information carriers compared to the alternatives (James and Ellington 1995:519–520). At the same time, having to face the problems of the RNA world mentioned above, these researchers suggest that the early self-replicators made of nucleotide units did not rely on the polymerization mechanism active in current cells. The present enzymatically catalyzed replication mechanism is based on the polymerization of single nucleotides on a template. It works only if the sugar residue in all the added monomers is pure ribose and all the nucleotides are of the right-handed kind. It is extremely difficult to envision that these conditions could have been met in the prebiotic situation. However, the earliest RNA molecules, say Kenneth James and Andrew Ellington of Indiana University, as well as other researchers, could have replicated by other methods, notably by serving as templates for the ligation, or joining together, of short oligonucleotides (James and Ellington 1995:521–526; Von Kiedrowski 1986; Li and Nicolaou 1994). Recent experiments in the laboratories of Orgel at the Salk Institute and Von Kiedrowski at Ruhr-Universität, Bochum, Germany, reveal several mechanisms for the ligation of two RNA segments aligned on a template (Zielinski and Orgel 1987). For instance, two segments, each consisting of three nucleotides, attach to an oligonucleotide template of six monomers, and then combine to form a complementary hexamer, a six-unit polymer. Each of the two ligated segments is complementary to part of the original template (Sievers and Von Kiedrowski 1994). In a similar manner, a template of six nucleotides can bind three segments of one, two, and three nucleotides. It is also possible to bind and ligate on a hexamer template segments of four and five nucleotides each. This procedure leads to elongation of the original hexamer template (James and Ellington 1995:526; Von Kiedrowski 1986).

It was found in these experiments that, in distinction to the known template-directed elongation mechanism, which relies on the addition of single nucleotides, template-directed ligation is much less sensitive to the purity of the sugar component and its homochirality, and to the "correctness" of the chemical bonds between the nucleotides. Sequences containing a variety of sugars, both left- and right-handed, and non-natural bonds between nucleotides could still bind and ligate on a template (James and Ellington 1995:521; Kanavarioti 1994:487). This is due in part to the fact that oligomers bind to templates more strongly than do monomers (Ferris 1994). James and Ellington believe that short oligonucleotides made of three to six residues and containing a variety of sugars and enantiomers could have accumulated in the prebiotic environment. Particular sequences in this mixed-backbone population could have then replicated via the ligation of shorter oligomers (James and Ellingon 1995:521). Moreover, several factors could have led to the evolution of the mixed population. Pure segments, in terms of constitution and chirality, bind to their templates more efficiently. Double helices containing only the correct components and linkages are usually more stable than helices containing mixed backbones and alternative chemical bonds. Therefore, a selection process over time could have led to the enrichment of the population with the correct products (James and Ellington 1995:524). Advocates of this approach claim that template-directed chemistry, especially template-directed ligation, could have functioned in the prebiotic environment as a tool to create a large population of polymers, out of which RNA catalysts could have emerged. The turnover of these polymers—that is, their rates of synthesis and decomposition—depended on their physical and chemical properties. It is out of this heterogenous pool that the RNA world eventually emerged (James and Ellington 1995; Kanavarioti 1994).

In an in vitro evolution experiment, Jack Szostak and David

Bartel, looking for ribozymes with new catalytic functions, were able to isolate a ribozyme with a copying-via-ligation function. Out of a large pool of random-sequence RNA molecules, a ribozyme was selected that functions as a template and also catalyzes the ligation of two complementary shorter segments aligned on the template (Bartel and Szostak 1993). The successful attempt to evolve ribozymes such as these lends credence to a hypothesis discussed previously, according to which the first major step toward an RNA world would have been the creation of a large pool, or "library," of different RNA-like polymers, out of which active ribozymes could eventually emerge. It should be pointed out that this is an optimistic scenario, which tends to disregard the fact that template-directed ligation experiments use, for various reasons, DNA rather than RNA sequences. More important, these experiments depend on tight control by the chemist over the various stages of the reaction. Furthermore, the question of the prebiotic synthesis of the building blocks, ribonucleotides, is not yet resolved (Ferris 1994:185). Realizing the enormous hiatus between in vitro experiments and the prebiotic scene, Bartel and Szostak comment that a realistic scenario for the prebiotic emergence of a ribozyme acting as a replicase from a primordial sequence pool that was truly random is hard to envision. They and other researchers believe that there is a way out of the dilemma, which involves minerals and metal ions acting as catalysts and thus contributing to the synthesis of non-random sequences (Bartel and Szostak 1993:1417).

THE ROLE OF CATALYSIS IN THE RNA-WORLD SCENARIO

Minerals, and also metal ions, are thought to have functioned prebiotically both as adsorbing surfaces and as catalysts. In experiments performed by Ferris, Orgel, and their colleagues, mentioned in chapter 10, the common montmorillonite clay catalyzed the formation of long oligonucleotides, containing as many as fifty monomers (Ferris et al. 1996). We have already noted that a certain minimal length and a definite chain structure are among the prerequisites for an active ribozyme. The mineral-catalyzed product would meet the minimal-length requirement. Catalysis by minerals or by metal ions could have also led to the synthesis of oligomers with a more definite structure. The general notion of the oft-mentioned optimistic scenario is that for a replicase ribozyme to evolve, a large pool of RNA and RNA-like sequences had to accumulate under prebiotic conditions. According to recent experiments, minerals could also have provided a library of surfaces on which heterogenous sequences could not only have been created but also copied. Ertem and Ferris studied certain oligonucleotides, containing both correct and incorrect bonds between their nucleotides units and also a mixture of linear and cyclic chains, that were synthesized on montmorillonite clay. They found that these heterogenous polymers can serve, despite their irregularity, as templates for the formation of a complementary copy. This template-directed copying performed on a mineral surface could also have increased the diversity of the pool of RNA-like sequences. The more diverse the pool became, the greater were the chances for the emergence of an RNA sequence able to catalyze its own copying, thus giving rise to a process of autocatalysis and to the development of an RNA world (Ertem and Ferris 1996).

Orgel believes that the development of RNA molecules on the primordial Earth was the watershed event in the development of life (Orgel 1994:61). However, as already noted, both he and Joyce regard the optimistic RNA-world scenario as unrealistic. According to Orgel, even most alternatives designed as pre-RNA molecules, such as the PNA polymer, are much too complex to be considered as the first genetic systems to arise (Orgel 1998:494). Orgel's group at the Salk Institute and Arrhenius's at the Scripps Institution for Oceanography are attempting to construct simpler pre-RNA molecules on the surfaces of minerals (Cohen 1995b:1926). The conception currently accepted by most researchers, optimists and pessimists alike, working within the context of the RNA-world conception is that catalysis of some sort was necessary for all stages leading to the emergence of life. Not only metal ions, minerals, and oligonucleotides, but also peptides, protein-like short chains, are invoked as active catalysts in the prebiotic environment. We have already noticed that Eigen's model relies on Fox's proteinoids as conceivable primordial enzymes preparing the ground for the emergence of the quasi-species and the hypercycle. In a similar manner, many current models, such as those of Noam Lahav of the Hebrew University (Lahav and Nir 1997) and Anastassia Kanavarioti of the University of California at Santa Cruz (Kanavarioti 1994:489–492), refer to peptide catalysts as being responsible for processes of template-directed replication.

In conclusion, a more general comment is in place. Traditionally, the genetic conception was associated with the assumption

that the highly improbable synthesis of a single self-replicating sequence was enough to start the emergence of life. Scenarios based
on this assumption tended to ignore the more deterministic aspects
of origin-of-life processes. This approach conflicted with the various metabolic-proteinic models, which insisted on certain physicochemical pathways or "channels" that led to the emergence of
organized metabolic systems. The current extensive research in prebiotic chemistry has resulted in the realization of the severe limitations imposed on any possible emergent system by the early
physical and chemical environment. The role allotted to contingent events in current RNA-world scenarios is still considerable.
Kanavarioti, for instance, speaks of diversification processes within
the population of primitive catalysts that led "not by selection but
by chance" to the emergence of better catalysts (Kanavarioti
1994:492). Yet the general feeling, as expressed aptly by James and
Ellington, is that "the strictures of prebiotic chemistry may have
narrowly defined the first self-replicating organic molecules" (James
and Ellington 1995:528). The realization of the narrow constraints
imposed on emerging life by the prebiotic environment has led to
a more open attitude toward non-genetic models. Both Joyce and
Eschenmoser, leading researchers of the chemistry of RNA, consider
the possible role of metabolic, non-genetic systems in changing the
chemical environment and thereby influencing the chances of the
emergence of a genetic system (Joyce 1989:222; Eschenmoser
1994:393–394). Despite the many theoretical and empirical differences between the genetic and the metabolic camps, current research is reshaping the traditional dividing lines.

12

AND YET,
METABOLISM

The hypothesis of a primordial RNA world, in one version or another, is highly popular among origin-of-life researchers, despite all
the difficulties involved. The ability of RNA molecules to function
both as carriers of information and as enzymes grants them a clear
advantage over protein molecules as possible "first comers" on the
ancient Earth. Even those critics of the RNA-world theory who claim
that another chemical world must have preceded the RNA world
still think that some genetic system, capable of self-replication, was
the first step on the way to life. According to this conception, it
was only through processes of replication and mutation that natural selection, and hence evolution, became possible. Though this
is the dominant trend in the field, it is by no means the only one.
The alternative "biochemical" or "metabolic" approach, which characterizes the first living system as an organized whole, performing
various catalyzed chemical reactions, is not dead at all. In fact, there
are some signs of its current renaissance.

In this chapter I will examine several metabolically oriented
origin-of-life theories. The basic assumption of all these theoretical attempts is that some form of a metabolic cycle emerged prior
to the appearance of a primitive genome and that the first living
systems could grow and evolve without genetic self-replication. I
will discuss the common elements of these theories and their dis-

tinctive features and point out the main differences between the metabolic and genetic approaches.

FREEMAN DYSON: THE "DOUBLE-ORIGIN" HYPOTHESIS

In the 1985 Tarner Lectures delivered at Trinity College, Cambridge, Freeman Dyson chose the origin of life as his theme because, in his judgment, this question has now become experimentally accessible (Dyson 1985:vii, 3). This was not the case, he pointed out, in 1943, when another physicist, Erwin Schrödinger, delivered a series of lectures entitled *What Is Life?* at Trinity College, Dublin (Schrödinger 1948). The main question Schrödinger dealt with was the physical basis of biological replication (Dyson 1985:2–3). Noting that Schrödinger set the path for future developments in molecular biology, Dyson nevertheless claimed that he neglected to ask some very important biological questions. Schrödinger focused his attention on the bacteriophage, a virus, that attacks a bacterium cell and can survive only within this cell, replicating its genome by exploiting all the biochemical machinery of the host bacterium. Due to the particular nature of the bacteriophage, Schrödinger could concentrate on its replicating functions. According to Dyson, Schrödinger put all the emphasis on the phenomena of replication, while almost ignoring the phenomena of organic metabolism, that is, the system of concatenated biochemical pathways carried out by protein enzymes (Dyson 1985:4–5). In his examination of Schrödinger's conception of a living system, Dyson restates the analogy drawn by the mathematician John Von Neumann between living organisms and the functioning of electronic computers, or mechanical automata in general (Von Neuman 1961–1963[1948]). Contrary to Schrödinger's focus, Von Neuman's logical analysis led him to claim that life is not one thing but two, metabolism and replication.

In the later-to-be-developed computer terminology, these stand, respectively, for hardware, which processes information, and software, which embodies it. Though allowing that both components are needed for a self-replicating automaton, Von Neuman maintained that, logically, hardware precedes software. Dyson is interested in applying Von Neuman's logical analysis to the origin-of-life situation. An outsider to the field who is not engaged in its research, he nevertheless feels that analyzing the "components of life" might shed light on the intractable origin-of-life problem. Von Neuman's insight was that logically, an organism made only of "hardware" can exist and maintain its own metabolism as long as it has enough food. (The question whether such an organism can evolve will be addressed later by Dyson.) On the other hand, an organism made of "software" with no "hardware" (the bacteriophage comes close to this) can survive only if other "hardware organisms" already exist, for it is by necessity a parasite, dependent on their metabolism for its existence and replication. Dyson wishes to correct the bias toward replication in molecular biology and origin-of-life research, where, he claims, it is often assumed that the origin of life was identical with the origin of replication. To correct this bias, Dyson makes an important distinction between replication and reproduction. Whereas in modern organisms the reproduction of a cell—its division into two daughter cells—depends on the replication of its genetic material, the two processes, he claims, could have been separate in the past. In Dyson's scenario, shortly to be described, the reproduction of a primitive cell preceded the exact replication of molecules, which developed at a later stage (Dyson 1985:5–9).

On the basis of his analysis of the two essential elements of life, replication and metabolism, and their relationship, Dyson suggests a double-origin hypothesis of the origin of life, which he describes as a combination of Oparin's and Eigen's theories. It will soon become obvious that Dyson's conception of the origin of life is, first and foremost, in agreement with Oparin's basic tenets. The ideas of Sidney Fox contribute also to Dyson's hypothesis. In the "first beginning," Dyson postulates, a primitive cell emerged— similar to the coacervates fashioned by Oparin—containing a metabolic system directed by protein-like enzymes. These cells could grow and divide, manifesting a primitive form of heredity, and they gradually developed a more efficient metabolism. Making use of this metabolism, a "second beginning" occurred in which nucleotides were synthesized; then a nucleic-acid polymer, an RNA molecule, appeared, and replication began. Dyson describes the emergence of the genetic machinery within the preexisting metabolic unit in terms borrowed from the University of Massachusetts biologist Lynn Margulis. Margulis suggested that major transformations in cellular complexity during evolution were due to the invasion of an existing cell by a smaller bacterium cell and to the subsequent symbiosis between the two entities (Margulis 1981). Dyson postulates that the RNA entity that emerged in the second beginning was such an invader of the preexistent metabolic system. At first it acted as "a parasitic disease" within the "infected cell." Gradually, after suffering casualties, infected cells learned to tolerate the RNA parasite, which then became a symbiont and finally an essential component of the cell (Dyson 1985:4–18).

To account for the first beginning, Dyson develops a mathematical "toy model" that incorporates the basic principles of the Oparin theory, but he does not go into the details of prebiotic chemistry (Dyson 1982; 1985:40–52). He describes the behavior of a population of polymers that might have been similar to the peptides known to us today, contained in a compartment. There is an external supply of new monomers, amino acids, and of energy needed for the reactions between monomers and polymers. In line with Motoo Kimura's neutral theory of molecular evolution (Kimura 1983), Dyson postulates that, following chemical reactions within the compartment, the population of polymers changed via random drift, that is, via statistical fluctuations. For the sake of simplicity, Dyson reduces all chemical processes among the polymers to a single reaction in which each polymer is engaged in replacing one monomer in another polymer. These substitutions are described by him as "mutations" that cause random changes in the population. A crucial assumption in Dyson's model is that certain amino acids substituted in certain sites on the polymer are "active": their insertion turns the mutated polymer into an enzyme. The enzymatic activity of a polymer consists in its ability to bring about more desired mutations, by inserting more active amino acids into active sites in other polymers. Relying on an intricate set of mathematical assumptions, Dyson claims to show that, under certain conditions, such a population of polymers is capable of "jumping" from its original disordered and inactive molecular state to a state of organized complexity, manifesting active metabolism. Active metabolism is achieved when "the cyclic shuffling [of monomers] maintains the active monomers at a self-sustaining high level" (Dyson 1985:46).

As long as cells have not reached this organized state, all changes in the population are caused by random drift. Once a population is engaged in active metabolism, however, it can incorporate new monomers from the outside and grow. This growth, Dyson contends, could bring about the physical division of the cell. If both daughter cells contain enough enzymes to remain in an actively organized state, the process of growth and division continues until the external supply of monomers is exhausted and some of the cells die. At this stage natural selection enters into the picture, as cells compete with each other for survival and have a better chance to survive if they learn to grow and divide more efficiently (Dyson 1985:67–68). Dyson's claim is that natural selection does not depend on the exact replication of a genetic molecule. In fact, according to his highly unorthodox definition of genetic information, the population of enzymes engaged in integrated, active metabolism is the carrier of genetic information. At this stage, he asserts, genetic information resides not in the individual components but in the architecture of the whole system (Dyson 1985:68–69). We are reminded here of Oparin's and Fox's claims that heredity depended at this early stage on the transfer between parent and daughter cells of a certain mode of organization manifested by a population of proteinic enzymes. It is Dyson's contention that natural selection among the metabolically active cells generated diversity and more sophisticated metabolic pathways. Among the products of this developed metabolism were the building blocks of nucleic acids, and later RNA polymers, which ushered in the second beginning described above (Dyson 1985:69–70).

When discussing Oparin's and Fox's models, I mentioned that both were criticized on the ground that without a self-replicating system evolution could not have occurred, and that therefore their scenarios lead to an evolutionary dead end. The critics assumed that there is no alternative to the mechanism of self-replication of genetic material. This mechanism guarantees continuity and at the same time, through the generation of new mutants, provides the raw material for natural selection (Orgel 1973:152; Eigen 1992:13–16). We have seen that Dyson faces this challenge by distinguishing between replication of molecules and the reproduction of primitive cells and by claiming that natural selection could have functioned through the transmission of the chemical constituents of his system from parent cells to divided daughter cells. Thus nucleic acids need not be the only possible source of genetic information. I will have more to say on this topic later in this chapter, in my discussion of other metabolic models that try to grapple with the same problem.

Dyson has been criticized for several of his key assumptions, and the validity of his mathematical model has been called into question (Lifson 1997:7). It should also be pointed out that developments in origin-of-life research since Dyson delivered his lectures have cast doubt on several of his empirical and theoretical notions. The primordial atmosphere is currently not regarded as of a highly reducing character, and the time window during which life emerged is estimated to be much shorter than Dyson assumed. The primordial soup scenario is doubted today by many researchers (Dyson 1985:20, 22, 29–30). More important, the distinction between the replication and metabolism approaches is not as sharply delineated today as it was a few years ago. For instance, Eigen, Dyson's "replication representative," relies in several of his publications on the catalytic activity of primitive enzymes, including Fox's proteinoids, which had a good chance of arising prior to RNA polymers and might have catalyzed the synthesis of such polymers. Earlier ideas about a "naked gene" have been forsaken, and Eigen sees the need for the hypercycle to be enclosed in a compartment as soon as it emerged. Furthermore, the discovery of ribozymes strengthened the RNA-world theory but at the same time sharpened researchers' awareness of the prebiotic difficulties involved in this theory. And yet the division between geneticists and metabolists has not disappeared, and Dyson's contribution to the understanding of the origin-of-life problem lies mainly in his philosophical perception of this division. More explicitly than most other models of the origin of life, his points to the connection between one's philosophical assessment of the nature of life and one's ideas about life's emergence.

HOMEOSTASIS—KEY TO LIFE

It is Dyson's conviction that the essential characteristic of each living cell is homeostasis, the ability to maintain a stable internal chemical balance in a changing environment (Dyson 1985:60–62, 71–72). Such a balance, without which no ordered metabolism is possible, is achieved through mechanisms of chemical control and feedback cycles. According to Dyson, a prevailing phenomenon in living systems, and also in ecological, economic, and cultural systems, is that efficient homeostatic mechanisms are complicated, consisting of a great variety of elements. Thousands of biological species generally make up an ecological system. A bacterium cell, in order to maintain homeostasis, has to include a few thousand different chemical components. Though it is reasonable to assume that the early mechanisms of life were simpler, Dyson believes that even a primitive living system had to maintain an internal chemical balance to survive in its environment, and hence was based on a complicated molecular structure. In his view, unlike the models based on a self-replicating RNA molecule, the Oparin model and his own satisfy the requirement for a heterogeneous multi-molecular system (Dyson 1985:62–64). Moreover, they also satisfy another of his essential criteria for a plausible origin-of-life model—they tolerate high error rates (Dyson 1985:73–74). Within the mathematical specifications of Dyson's model, he calculates that his system can function with an error rate as high as 25% (Dyson 1985:52). This stands in stark contrast to the strict control and accuracy required by Eigen's hypercycle model (Dyson 1985:36). Dyson is drawn to Darwin's image of life as a "tangled bank" of many biological species interacting with each other in a complex web (Dyson 1985:73). This error-tolerant image is the one he would like to adopt not only for ecological communities but also for the origin of life.

STUART KAUFFMAN: COMPLEXITY AND SELF-ORGANIZATION

The theoretical biologist Stuart Kauffman of the Santa Fe Institute has advanced an origin-of-life theory that is reminiscent in some respects of Dyson's hypothesis. Kauffman's approach to the problem forms part of his wider conception regarding the behavior of complex systems manifesting self-organization. During the last two decades a theoretical field has been studying phenomena of complexity in systems both natural and social (Pagels 1988; Goodwin and Saunders 1989; Stein 1989). Proponents of the "complexity theory," most of whom are members of the Santa Fe Institute, claim that under certain physicochemical conditions many systems are able to spontaneously self-organize into a highly ordered state. These systems consist of a large number of components, and the phase transition from disorder to order depends on intricate interactions among the different components (Mehta and Baker 1991).

   Such complex systems are thermodynamically open: they incorporate energy and matter from their external environment, thus maintaining a far-from-equilibrium thermodynamic state. Free exchange of matter and energy with the environment is harnessed to the building of a complex organization. As opposed to the general tendency in the physical universe and within closed systems toward maximum disorder, the Second Law of Thermodynamic in these open systems in fact favors the building of order and organization. This understanding refutes the traditional notion that highly organized systems, in particular living organisms, are incompatible with the second law. Underlying a thermodynamic approach to biological evolution is the claim that the second law has a central, positive role in evolution. The structuring and organization of biological systems are conceived as a means to build pathways for the dissipation of unusable energy and material (Weber et al. 1988). One of the proponents of this paradigm, Jeffrey S. Wicken of Pennsylvania State University, contends that "dissipation through structuring is an evolutionary first principle" (Wicken 1987:5). These ideas were inspired by the theoretical work of the physical chemist and Nobel laureate, Ilya Prigogine. He characterized certain physical systems as "dissipative structures," organized structures that form spontaneously from a disordered state, constructing and preserving their self-organization through the constant input of efficient external energy and the output of less efficient energy to the environment (Nicolis and Prigogine 1977).

One of the properties characterizing certain complex systems, such as weather systems, is "deterministic chaos," which became widely popularized through the "butterfly effect," the influence of very minor climate changes in one part of the globe on the weather on the other side of the globe. Though obeying mathematical equations and thus being defined as deterministic, chaotic systems nevertheless show extreme sensitivity to very small changes in initial conditions, and thus may pass from a relatively orderly state to a totally disordered and unpredictable one (Crutchfield et al. 1986; Gleick 1987).

Kauffman claims that some complex systems may manifest behavior he calls "antichaos": from a disordered state, these systems can spontaneously turn into an organized whole (Kauffman 1991:78). According to Kauffman, the ability of complex material systems in far-from-equilibrium conditions to self-organize, or as he puts it, to "crystallize," into a high degree of order is a crucial mechanism in the evolution of the living world. On the basis of this phenomenon, he insists, we can explain the origin of life, significant events in biological evolution, and also the processes of embryonic development and differentiation of cell types (Waldrop 1990). In Kauffman's view, the accepted opinion of evolutionary biologists that natural selection is the sole source of order in the living world is incorrect (Kauffman 1993:10, 16, 285). He perceives spontaneous self-organization, an innate property of certain complex systems, as a complementary factor in evolution, providing different patterns of order on which natural selection can act.

Complexity theory in general and its application to biological systems, in particular to the origin-of-life question, are so far based not on empirical data gained through laboratory experiments but on mathematical and computer models. Many scientists question the relevance of Kauffman's theory to real biological systems (Maynard Smith 1995; Joyce 1989), and more acrimonious critics describe his theory as losing contact with reality (Dover 1993). Yet Kauffman maintains that on the basis of his mathematical models one can make predictions that are consistent with observed biological facts (Waldrop 1990:1543; Emmeche 1994:104–109).

Kauffman's theory of the origin of life postulates as a first step of organization the emergence of a self-reproducing metabolic system consisting of interacting catalytic polymers. The key concept in the theory, only implicitly alluded to by Dyson, is autocatalysis. Whereas a self-replicating molecule, such as an RNA polymer, is a single autocatalytic unit, doubling itself in each replication cycle, Kauffman describes an autocatalytic set of catalytic polymers, in which no single molecule reproduces itself but the system as a whole does (Kauffman 1993:285–288). It should be noted that Kauffman assumes his sets of polymers to include catalytic peptides and catalytic RNA molecules (ribozymes). Relying on previous work done by origin-of-life experimentalists and theorists in both the metabolic and genetic traditions, he points out the feasibility of the prebiotic synthesis of peptides capable of catalyzing the formation and cleavage of peptide bonds and the feasibility of the prebiotic synthesis of RNA-like short sequences capable of catalyzing the ligation and cleavage of even shorter RNA sequences. Unlike some versions of the RNA-world theory, Kauffman's model of an autocatalytic set does not depend on a replicating ribozyme, whose prebiotic synthesis is estimated as highly improbable. The model also assumes, for the sake of efficiency, the encapsulation of the catalytic polymers in a closed volume, a sort of compartment like Oparin's coacervate, or Fox's microsphere. Kauffman also discusses the possibility of synthesis of catalytic polymers under conditions that favor the removal of water, such as the adsorption of the reacting molecules on a mineral surface. This would increase the likelihood of chemical bonds between amino acids or nucleotides, which in the presence of water tend rather to hydrolyze, or break (Kauffman 1993:298–300).

Postulating that all these preparatory conditions can be fulfilled prebiotically, Kauffman can advance his main idea, the core of his model. The goal is to describe how a collection of catalytic polymers undergoes a "phase transition" to a state of "catalytic closure" in which an organized set of polymers catalyzes its own reproduction. It is Kauffman's claim that this transition occurs spontaneously once a certain critical level of molecular diversity is reached (Kauffman 1993:309). More specifically, the model assumes for the sake of simplicity that peptides are composed of only two amino acids, a and b, and that the ligation and cleavage reactions in the set are of the types ab + b $\leftrightarrows$ abb or baabbbab $\leftrightarrows$ bab + baabb, including also exchanges like aaba + bbbb $\leftrightarrows$ aabb + a + bbb. Kauffman's assumptions as to the composition of the ribozymes in a collection from which an autocatalytic set might arise are more complicated, but are nevertheless compatible, he claims, with many recent experiments on the template-directed ligation of single-stranded RNA molecules. Kauffman contends that as the maximum length of the polymers increases, the number of possible polymers increases and the number of reactions among these polymers increases yet faster. Eventually, when the number of reactions is high enough relative to the number of polymers, a stage is reached where the last step in the synthesis of almost all polymers is catalyzed by another polymer in the set. At this point the whole set becomes collectively autocatalytic (Kauffman 1993:309, 323; Maynard Smith and Szathmáry 1995:67–72).

Several critics have argued that the formation of an autocatalytic set from a population of catalytic polymers is unlikely (Lifson 1997:6–7). Assuming that such metabolic cycles can arise under the conditions specified by Kauffman, his model has still to tackle the difficult question of evolution: Can the cycles evolve? Kauffman has to overcome the serious challenge facing any metabolic scenario, be it Oparin's, Fox's, or Dyson's: he has to account for a mechanism of evolution in a system lacking a genome. Whereas both heredity and variations are inherent to a system containing a replicating genetic molecule, what is the source of heritable variations in Kauffman's system? We have seen that Dyson postulated the reproduction of cells via division into daughter cells and defined "genetic information" as the architecture, or ordered pattern, of protoenzymes transmitted in each cell division to the new generation of cells. Relying on the vague notion of statistical fluctuations, he did not explicitly discuss the question of variations or mutations. Kauffman suggests several mechanisms that could plausibly explain how the metabolic set could produce mutant polymers and new autocatalytic sets on which natural selection could act. He claims that in addition to catalyzed reactions in the autocatalytic set, spontaneous reactions utilizing components of the set also occur, generating a "shadow set." Within this shadow set, mutant polymers arise, which under certain conditions may be added to the initial system. For that to happen, a collection of such mutant products must jointly catalyze their own formation from one another and from the core system. When this occurs, another "metabolic loop" is added to the original set, which has now become a new autocatalytic set. At the same time, if one of the spontaneously produced molecules happens to inhibit one of the catalyzed reactions, the original set will lose a metabolic step and will thus be changed (Kauffman, 1993:330–333).

One of the strongest lines of criticism of such a mechanism of evolution was made by Eigen, who already in 1971 had examined the question of whether an autocatalytic set of peptides would be able to evolve (Eigen 1971:498–503; Maynard Smith and Szathmáry 1995:68, 71). Eigen pointed out that when a mutation occurs in one catalyst in the set, the whole cycle would retain its autocatalytic nature and reproduce only if all the other catalysts mutated in a synchronized fashion. Since this condition is highly improbable, Eigen rejected the option and adhered to the idea that heredity, mutation, and evolution could be attained only through the self-replication of nucleic acids. Without mentioning Eigen's name, Kauffman acknowledges this problem when he states that only sets that achieve collective catalytic closure can survive, admitting that even among infinite numbers of polymers the number of such sets will be finite (Kauffman 1993:332). However, he describes another, Dyson- or Oparin-like, source of heritable variations, based on the ability of the compartments containing the autocatalytic sets to divide (Kauffman 1993:331–332). Upon each division, he says, there is a random distribution of the original content between the two daughter cells. Since not all the original catalytic polymers will be equally divided, such stochastic differences could account for the evolution of new autocatalytic sets. Under given selection conditions, for instance, when a specific variety of molecules in the environment serves as "food" for the organizing systems, some sets will incorporate this food more efficiently, will reproduce faster, and will be more fitted to their environment. These mechanisms of evolution do not seem to satisfy Kauffman's critics, who contend that, at best, he can provide only a limited range of hereditary variations (Maynard Smith and Szathmáry 1995:71). Notwithstanding this criticism, Kauffman elaborates on the further development of his system, describing how replicating nucleic acids can eventually develop within the metabolic framework of the evolving autocatalytic sets (Kauffman 1993:357–366).

I have already commented on the skeptical attitude of many biologists toward the relevance of any mathematically idealized system to actual biological systems. And yet there is no doubt that Kauffman's model focuses on one of the crucial philosophical questions pertaining to the origin of biological systems—the uniqueness of biological organization. The most noticeable characteristic of this organization, he contends, is the close interdependence of its different components that collaborate to assure the functionality of the whole system (Kauffman 1993:16, 21–22). I have already examined the phenomenon of biological vicious circles in my discussion of the chicken-and-egg problem. Not only is the biological whole dependent on the existence and activity of its components, but each of the components depends on all the others and on the system as a whole. Such purposeful, closely knit interdependence is not found in other natural objects, but only in man-made artifacts designed according to a definite plan and for a definite purpose. It is thus no wonder that the living organism is described in terms of design by both creationists, who believe in a God-created design, and by molecular biologists, who speak of a genetic program developed during the evolutionary process. I will devote part of my discussion in the next chapter to philosophical issues involved in the question of the emergence of biological design.

It is important to note that Kauffman wishes to distinguish his ideas from the philosophical approach, which he associates with "replication-first" theories (the genetic tradition in my terms), according to which evolved organisms are contingent contraptions resulting from historical accidents (Kauffman 1993:13; Waldrop 1990:1543). In his view, not only are living organisms complex wholes, they also emerged in the holistic fashion demonstrated by his model (Kauffman 1993:287–288, 21–22). Like Dyson, he regards life as depending, from the very start, on a certain level of complexity (Kauffman 1993:294), and he wishes to account for this initial complexity without regarding it as a historical accident. Kauffman believes that unlike the various RNA-replication theories, his conception of the origin and evolution of life is compatible with the basic features of the living organism. Spontaneous self-organization, he claims, which brings about the emergence of a complex whole, is a basic physical principle manifested in certain systems. According to Kauffman's metabolic theory, life is "an emergent collective property of complex systems of polymer catalysts" (Kauffman 1993:287). Moreover, the processes responsible for self-organization in such systems are fast and highly probable, and hence are compatible with recent estimates of the short time window during which life emerged. Kauffman believes that life is a universal phenomenon. Given the appropriate physical and chemical conditions on other planets, the emergence of life is to be expected (Kauffman 1993:16, 22, 330).

WÄCHTERSHÄUSER'S THEORY: THE EMERGENCE OF LIFE ON CRYSTALS OF "FOOL'S GOLD"

Unlike the mathematical models of Dyson and Kauffman, a few other metabolism theories are based on detailed chemical and biochemical processes, referring to specific characteristics of prebiotic chemistry. Among these theories, one of the most original and extensive was developed recently by the German organic chemist Günter Wächtershäuser. Wächtershäuser's biography is quite unusual. Following training as an organic chemist, he became a lawyer specializing in patent law. His consuming interest in the problem of the origin of life led to his first articles on the subject, which were published in the 1980s with the encouragement of the renowned philosopher of science Karl R. Popper. Wächtershäuser professes to have based the methodology of his investigations on the tenets of Popper's philosophy of science, especially the rejection of "the imagined need to derive biology from chemistry" (Wächtershäuser 1992:88). He claims that though the laws of chemistry constrain the myriad biological possibilities, a theory of the emergence of life, which he regards as equivalent to a theory of the emergence of biochemical pathways, is a biological theory (Wächtershäuser 1992:88). It is interesting to speculate whether Popper's encounter with Wächtershäuser's ideas could have changed the philosopher's attitude toward the origin-of-life problem. Previously, following Monod's contention that life emerged from inanimate matter by an extremely improbable combination of chance circumstances, Popper claimed that the origin of life is "an impenetrable barrier to science and a residue to all attempts to reduce biology to chemistry and physics" (Popper 1974:270). Accepting an emergence-of-life theory, be it Wächtershäuser's or someone else's, would amount to an acknowledgment that the origin of life is not impenetrable to science and that there is a way to bridge the gap between chemistry and biology without having to reduce the latter to the former.

Wächtershäuser rejects most of the basic assumptions underlying other origin-of-life models, and his theory thus forms a category of its own. All the soup theories suggested so far—both metabolic and genetic—assume that the first biological systems depended on taking up food, in the form of organic compounds, from their environment. These basic organic compounds were supposed to be made from simple inorganic constituents. There is thus a common assumption that emerging life was heterotrophic, that is, it got its food from an external source (*hetero*—other in Greek; *trophe*—food). Wächtershäuser, on the other hand, contends that the first organism was an autotroph, self-feeding—capable of synthesizing all its organic carbon constituents from carbon dioxide (CO_2) or from other simple inorganic carbon molecules. Autotrophy is very common among extant organisms, the most familiar example being plants, which synthesize their food from carbon dioxide and water under the influence of solar energy. According to Wächtershäuser, there are two major reasons why all other origin-of-life theories deny the possibility of an autotrophic origin. First, postulating that various organic substances were synthesized in the prebiotic environment, different syntheses can be imagined to have taken place under diverse chemical conditions and in different locales, for instance in meteorites or in the deep ocean. An autotrophic origin, on the other hand, is restricted, in Wächtershäuser's words, to a "single theater" where everything is supposed to have happened. Second, whereas organic synthesis in the soup might work with all sources of energy, it is difficult to envision a geochemically plausible energy source that could be linked directly to carbon dioxide within the organism, transforming it into organic compounds (Wächtershäuser 1992:89–90). It will be presently shown how Wächtershäuser's scenario handles these two problems.

On the basis of available data on the geochemistry of the primordial Earth, Wächtershäuser asserts that organic compounds could not have been formed in the soup. Relying on considerations pertaining to the stability of the organic products and the rate of chemical reactions taking place in solution, he concludes that the chances for getting prebiotic products in the primordial soup were extremely slim: under such circumstances reactions of decomposition are favored over composition. This understanding leads him to the idea that all the substances required for the emergence of life were produced not in solution but on the solid surface of a mineral. Prebiotic reactions had to be "two-dimensional surface reactions," which differ in principle from processes that occur in a solution (Wächtershäuser 1992:94–95). As we have already seen, the notion of mineral surfaces as aids in polymerization reactions has been considered by other researchers, but Wächtershäuser's theory differs from both "conservative" and "radical" mineral theories in its emphasis on the autotrophic nature of the first organisms.

Wächtershäuser has a specific mineral in mind as both surface and participant in the emergence of life. Organic compounds formed, he suggests, on the surface of pyrite (FeS_2), a mineral made of iron and sulfur that is sometimes called "fool's gold." (Prospectors mistook it for the coveted metal because of its characteristic glimmer.) Pyrite is a ubiquitous mineral found even in the oldest sedimentary rocks. It is the most stable iron mineral under anaerobic conditions. Moreover, the chimneys of hydrothermal vents consist largely of pyrite, one fact among many that make Wächtershäuser's theory compatible with the hydrothermal-systems hypothesis for the origin of life. However, pyrite was chosen by Wächtershäuser not only for these geochemical reasons. It turns out that the common reaction in which pyrite is synthesized from hydrogen sulfide (H_2S) and an iron salt (FeS), both of which were widespread on the primordial Earth, leads to the release of energy and hydrogen:

$$FeS + H_2S \rightarrow FeS_2 + H_2 \ (+ \ energy).$$

The released hydrogen provides the reducing power that is needed for the synthesis of organic compounds out of carbon dioxide on the primordial Earth (Wächtershäuser 1992:91). This is Wächtershäuser's solution to the dilemma concerning the synthesis of organic compounds under a non-reducing atmosphere. The formation of pyrite solves another crucial problem. It provides a source of chemical energy on which an autotrophic origin of life depends. Wächtershäuser first considered the possibility of photo-energy and rejected it: visible light is not powerful enough to generate reducing power, and ultraviolet light, which is powerful enough, might destroy organic substances. The alternative is thus chemical energy, and with pyrite Wächtershäuser has found an appropriate candidate. The synthesis of pyrite serves as a source of chemical energy for the synthesis of organic compounds taking place on its surface. By providing such a surface for synthesis, pyrite also satisfies the condition of restricted locality. Wächtershäuser called all the reactions that depend on the reducing power released through the synthesis of pyrite "pyrite-pulled reactions" and his theory of the origin of life is thus a theory of "a pyrite-pulled chemo-auto-origin" (Wächtershäuser 1992:91).

Pyrite has another crucial advantage: its surface has positive electrical charges that can bind to negatively charged chemical groups. In this manner, the negatively charged form of carbon dioxide (HCO_3^-) can be bonded to the pyrite surface. Moreover, the organic products synthesized from carbon dioxide under the reducing power of the released hydrogen become bonded to the pyrite surface *in statu nascendi*, that is, while they are being formed. This important point should be reiterated. Wächtershäuser is describing here an intricate process whose components are tightly connected: the synthesis of pyrite from its constituents, the release of hydrogen and chemical energy during this synthesis, the bonding of carbon dioxide to the produced pyrite, and the fixation of the bound carbon dioxide to organic compounds, which become attached to the surface while being formed. The various bonded organic compounds continue to accumulate and undergo subsequent reactions while still attached to the surface.

It is Wächtershäuser's claim that the first stage in the emergence of life was the development of such a surface organism, whose different constituents established a primitive metabolism on the pyrite surface. This initial chemo-autotrophic organism, called by Wächtershäuser a "surface metabolist", is defined by him as "the totality of all surface-bonded organic constituents and of their pyrite base" (Wächtershäuser 1992:96). The environment is defined as "the totality of the non-surface-bonded molecules in the water phase. These serve as sources of inorganic nutrients and the water phase also serves as a sink for detached organic products of decay." Surface metabolism, consisting of transformations of the different chemical groups and their interactions, is made possible by the migration or diffusion of molecules on the mineral, to which they are continuously bonded. According to Wächtershäuser, the mechanism of two-dimensional diffusion on a surface assures the lateral migration even of large negatively charged molecular structures, which may detach locally from the surface but still retain their cooperative overall bonding. It should be pointed out that the surface metabolist has no definite boundaries on the mineral, and its organic constituents can spread into vacant pyrite crystals (Wächtershäuser 1992:96).

Wächtershäuser's theory of the emergence of metabolism is in fact a theory of the emergence of early biochemical pathways, which he regards as very much in need. He deplores the fact that the triumph of biochemistry in explaining the central biochemical cycles in extant organisms is not being matched by an understanding of the early evolution of these cycles. Thus he suggests a "methodology of retrodiction," a combination of chemical and evolutionary explanations that will attempt "to draw lines of descent from the extant organisms back towards the first reproducing metabolist" (Wächtershäuser 1992:121). Ultimately, such explanations will show how all the pathways of current central metabolism, which are common throughout the living world, can be "retrodictively transformed into archaic pathways" (Wächtershäuser 1992:88). Furthermore, he claims that the point of convergence of all these pathways is pyrite-pulled metabolism. Due to limitations of space, I will not describe Wächtershäuser's detailed methodology of biochemical retrodiction. He foresees the development of experimental testing of his postulates as ultimately leading to a "historical ordered table of biochemistry." I will presently comment on several recent experimental results that seem to confirm some of Wächtershäuser's basic postulates.

Empirical support for Wächtershäuser's theory can be drawn from the fact that the pyrite-pulled chemo-auto-origin is highly compatible with the submarine scenario and its geochemical setting of the origin of life on Earth as currently perceived by quite a few researchers (Vogel 1998a). Discussing the mechanisms of surface metabolism, Wächtershäuser indicates that they are favored by high temperatures and atmospheric pressure: the surface metabolist is both thermophilic and barophilic. The positive dependence on increasing temperatures is reinforced by acidic conditions. Citing recent investigations of deep-sea vents and their environment, Wächtershäuser thus concludes that "the chemical conditions most favorable for a pyrite-pulled chemo-auto-origin (elevated temperatures, high pressure, a nearly neutral pH) all happen to be geochemically most plausible" (Wächtershäuser 1992:93). He therefore suggests that life on Earth might have emerged at volcanic or hydrothermal sites. These conditions are also in agreement with the conclusion Woese draws from his molecular studies of the tree of evolution. Since the most ancient microorganisms among the Archaea and Bacteria are thermophilic, life could have originated under thermophilic conditions. (I have discussed in chapter 10 ideas that question this conclusion and stress the need to distinguish between common ancestor and emerging living systems. Wächtershäuser's and Woese's ideas, on the other hand, emphasize the continuous nature of these evolutionary stages.) It is important also to point out that the microorganisms that belong to the Archaea group are autotrophs: like Wächtershäuser's conjectured metabolists, they produce their organic constituents from inorganic nutrients in their external environment (Gray 1996:300). Interestingly, the last common ancestor, according to Woese, also engaged in sulfur metabolism (Woese 1987).

Whereas all the other metabolic theories discussed so far—Oparin's, Fox's, Dyson's, and Kauffman's—assume the emergence of metabolic cycles within a protocell, Wächtershäuser's conjectured first organism, being a surface metabolist, is a non-cellular system. Wächtershäuser claims that the basic characteristics of the surface metabolist lead in a continuous way to a process of self-cellularization. This depends on the simultaneous occurrence of two processes: the ongoing formation of pyrite crystals with the release of reductive power; and the fixation of carbon dioxide, first reduced to methane (CH_4), into long organic polymers that bond to the pyrite crystals. These organic polymers, made of $-(CH_2)-$ units, are polar lipids that bind strongly to the pyrite, pushing the water away from the surface. This protective, water-repellent layer gradually becomes a membrane made of inner hydrophobic and outer hydrophilic molecules. In this water-protected environment, certain organic constituents, among them polypeptides and nucleic acids, tend to form more easily on the pyrite and are protected by the lipid membrane against detachment from the surface (Wächtershäuser 1992: 104–106).

In Wächtershäuser's scenario, the growth of the newly formed discrete pyrite crystals and the simultaneous accumulation of lipids on their surface finally lead to the formation of a closed membranous envelope around the pyrite grains. The scenario now relies on the phenomenon of a "second nucleation," by which a pre-existing pyrite crystal "buds" into a new crystal nucleus. On the surface of the new crystals, a new organic layer and lipid membrane start to form. This process can be seen, Wächtershäuser claims, as a "growth-related division" and "the evolutionary precursor of a cell division" (Wächtershäuser 1992:107). Among the further stages in the evolution of cellular organization, the next major one is the detachment of the lipid membrane, including the inner organic layer, from the pyrite grains. The detached membrane then serves as the envelope of a new cell, in which the metabolic pathways, after an interim period of development, can explore a pyrite-free, independent existence. The development of various mechanisms at the cell membrane state gradually enables the organism to utilize organic molecules in its environment, free of its previous dependence on pyrite formation as the sole source of energy. Unlike origin-of-life theories that start from a heterotrophic state, in Wächtershäuser's scenario this state is a late development of the initially autotrophic organism (Wächtershäuser 1992:104–110).

As in other metabolic theories, nucleic acids are latecomers to Wächtershäuser's scenario: "the invention of a metabolism and not . . . the inventors of a metabolism" (Wächtershäuser 1992:185). The theory thus has to include a mechanism of evolution that will account for reproduction and variations in the absence of nucleic acids. Wächtershäuser's mechanism of evolution depends on the basic characteristics of surface metabolism, the bonding of organic compounds to the pyrite as they form, and the concomitant growth of new pyrite crystals. The element of reproductive inheritance is added to this autotrophic growth by one additional assumption. Reproduction and inheritance result if among the surface-bonded organic molecules a self-sufficient chain of metabolic reactions develops in which, through the fixation of carbon dioxide, the initial set multiplies itself. Wächtershäuser is describing here a process of surface-bonded autocatalysis, which involves the multiplication, or doubling, of the constituents of a chemical set through the incorporation and reduction of carbon dioxide and the rearrangement of the products. The doubled constituents formed in the autocatalytic process become bonded to vacant pyrite surfaces. We have here a process of reproduction and inheritance due to the multiplication of organic constituents and the growth of new pyrite crystals to which these organic molecules bond (Wächtershäuser 1992:110–111). The reproductive character of this process is even more obvious when it is associated with a budding of new pyrite crystals and a sort of cell division.

It should be noted that the existence of pyrite-bonded autocatalytic cycles is to a great extent the outcome of another basic characteristic of the surface metabolist, which is undergoing processes of chemical selection due to the differential propensity of the various organic constituents for detachment from the pyrite surface. I have already mentioned that certain organic polymers, including peptides and nucleic acids, are strongly attached to the surface. It is also through the process of selection by detachment that the relatively few self-sufficient autocatalytic cycles are maintained on the surface (Wächtershäuser 1992:96–97, 110–111).

I have discussed Wächtershäuser's mechanism of reproduction and inheritance. Evolution, however, also depends on the possibility of variations. Wächtershäuser relies on the initiation of novel branch reactions, some of which are extensions of the autocatalytic cycles. Most of the metabolic novelties are transient and are quickly detached from the surface or decomposed. Yet these transient novelties, if they prove inheritable, can function as evolutionary variations. This happens when a branch constituent is able to catalyze one of the reactions in the autocatalytic cycle. If the new catalyst bonds to a new pyrite crystal with the reproduced products of the catalyzed autocatalytic cycle, it will become a true component of the inherited expanded cycle. Wächtershäuser indicates, however, that the chances for that to happen are very small taking into account the great propensity of branch constituents to be detached or to decompose. Greater chances for true inheritance of the new catalyst can be envisioned in the rare cases when it catalyzes not only a reaction in the autocatalytic cycle but also a reaction leading to its own production. Wächtershäuser postulates such rare but nevertheless possible occurrences on the basis of the fact that chemical reactions are class reactions and hence a novel constituent may catalyze more than one chemical reaction (Wächtershäuser 1992:111–112).

Wächtershäuser claims that these rare occasions or mutations, "low-propensity ignitions of autocatalytic cycles" (Wächtershäuser 1992:89), are the raw material of biochemical evolution. Though he is here allotting a crucial role to unique and improbable events in the origin of life, that by no means implies that he regards the emergence of life on Earth as a unique and unrepeatable event. On the contrary, the model suggests that conditions for an auto-origin might in fact continue to exist even today, and during the history of the Earth might have initiated repeated processes of evolution. In places that are suitable also for the growth of heterotrophs, the surface metabolists would be immediately devoured by extant heterotrophic microorganisms. (This was the basis for Darwin's, Oparin's, and Haldane's argument that life could have emerged only on a lifeless Earth.) However, in environments unsuitable for heterotrophs, surface metabolists could flourish today and continue their evolution (Wächtershäuser 1992:114). Moreover, Wächtershäuser points out that the hot temperatures required by the surface metabolists could create separate ecological niches in which autotrophic pyrite organisms could safely emerge and evolve (Wächtershäuser 1992:103).

Assuming that in the appropriate niches life could have emerged more than once, Wächtershäuser asks, how can the universality of biochemistry and the genetic code be explained? Why is there only one tree of life? As an answer, he considers two extreme evolutionary possibilities: alternative forms of life exist in special isolated chemical environments that have not been found yet, or a rampant lateral transfer of genes (or elements equivalent to genes) between organisms in different localities guaranteed one evolutionary development. A middle-ground situation is possible, says Wächtershäuser, if lateral transfer might have occurred but was kept in check through evolved mechanisms of biochemical isolation (Wächtershäuser 1992:114). It is noteworthy that among the reasons geochemists find the hydrothermal origin-of-life scenario attractive is that hydrothermal systems can be postulated "to have existed through the entire length of Earth's history" (Holm 1992:9). Under these conditions autotrophic life could have originated many times. However, because of geological changes throughout history, hydrothermal activity was probably more pronounced on the early Earth.

Our examination of Wächtershäuser's theory has revealed its rich chemical and biochemical content, its fertile approach in accounting for different aspects of the origin of life, and its consistent logical structure. Another crucial asset is the fact that the theory's different claims can be experimentally tested. Relying on his methodology of retrodiction, Wächtershäuser discusses a few examples of extant autocatalytic metabolic cycles that could have emerged and evolved according to his theory. Notable among them is the reductive citrate cycle (RCC), considered the oldest autocatalytic carbon-fixation cycle, which Wächtershäuser claims can be retrodicted to an archaic form, the pyrite-pulled formation of succinate from carbon dioxide (Wächtershäuser 1992:127). Several recent experiments exploring the connection between the reducing power produced in the synthesis of pyrite and a carbon-fixation cycle similar to the reductive citrate cycle seem to confirm Wächtershäuser's postulates (Huber and Wächtershäuser 1997:246). A specific prediction of Wächtershäuser was confirmed in another experiment, in which iron sulfide reacted with hydrogen sulfide and carbon dioxide to form methyl thiol (CH_3SH) (Heinen and Lauwers 1996). According to Wächtershäuser's theory, this compound serves as an intermediate, which upon the addition of carbon monoxide in the presence of iron-sulfur minerals and with the participation of nickel ions as catalysts forms an active compound with a C-C bond (Huber and Wächtershäuser 1997:245, 246). Indeed, Wächtershäuser and Claudia Huber of the Technische Universität in Munich reported on the formation from methyl thiol of such a compound, an activated form of acetic acid that resembles acetyl-coenzyme A, a major intermediate in some of the most important biochemical pathways. It should be noted that this active intermediate is a starting material for the synthesis of amino acids (Vogel 1998:628). Huber and Wächtershäuser performed their experiment at 100°C in the presence of an aqueous slurry of iron- and nickel-sulfide minerals and carbon-monoxide and carbon-dioxide gases, conditions prevailing near submarine volcanic-gas vents. It is significant that under these conditions, which are compatible with the geochemical setting of a pyrite-pulled origin, a chemical reaction was obtained that the researchers claim "can be considered as the primordial initiation reaction for a chemoautotrophic origin of life" (Huber and Wächtershäuser 1997:245).

Huber and Wächtershäuser performed another experiment modeling a similar hydrothermal setting in which amino acids were linked to form dipeptides and a few tripeptides. This experiment was also conducted at 100°C in the presence of nickel- and iron-sulfide minerals, which supposedly catalyzed the activation of amino acids by carbon monoxide and hydrogen sulfide as well as the linking together of the activated amino acids to form peptide bonds (Huber and Wächtershäuser 1998). As noted by researchers supporting the hydrothermal systems scenario, such as James Ferris and Norman Pace, unlike other experiments in the past in which amino acids were linked together to form peptides under implausible prebiotic conditions, Huber and Wächtershäuser used only reagents that are available in volcanic exhalations in a hydrothermal setting (Vogel 1998a:627–628). Other origin-of-life scientists, on the other hand, notice that the concentration of carbon monoxide used by Huber and Wächtershäuser in their experiment is not to be found in hydrothermal vents (Vogel 1998a:628; Schoonen et al. 1999). The fact that no synthesis of amino acids themselves under geothermal conditions has been obtained is noted by other critics. Wächtershäuser, however, is now in the process of trying to produce amino acids under similar hydrothermal conditions.

Many origin-of-life investigators have rejected the autotrophic option and its restriction to a very limited chemical environment (Oparin 1953:202–208; De Duve 1991:152). Many are highly critical of the hydrothermal scenario. Criticism has also been addressed to the specific model suggested by Wächtershäuser (De Duve and Miller 1991). Nevertheless, the possibility of a pyrite-pulled auto-origin cannot be lightly dismissed. Wächtershäuser's theory presents a coherent picture of the emergence of life, and most importantly, of the evolution of biochemical pathways. As such it is a strong alternative to the various versions of the RNA-world theory. Differing in its basic assumptions from the other metabolic theories discussed in this chapter, Wächtershäuser's model nevertheless indicates that a metabolic approach to the origin-of-life problem is an option that should be seriously considered. His pyrite theory belongs to a group of origin-of-life theories that focus on the role of iron-sulfide minerals and on the submarine environment. Citing theoretical considerations and experimental work, Michael Russell of the department of geology at the University of Glasgow and his colleagues postulate the emergence, under hydrothermal conditions, of "bubbles" and "fine chimneys" made of colloidal iron-sulfide membranes. These structures are supposed to have functioned as metabolizing protocells, synthesizing complex organic molecules with the help of chemical energy produced by the iron-sulfide system (Cairns-Smith et al. 1992; Kaschke et al. 1994). More will be said about this theory in the next chapter.

LIPID VESICLES—ANCESTORS OF CONTEMPORARY CELLS?

The first organism, according to Wächtershäuser, was a non-cellular surface metabolist, which only gradually, going through a semi-cellular stage, evolved into a membrane-engulfed protocell. Traditional genetic theories spoke of a "naked gene," postponing the emergence of a protocellular structure to a later evolutionary stage. Manfred Eigen, for instance, discussing the evolution of genetic information within the quasi-species population, pointed out that "organization into cells was surely postponed as long as possible. Anything that interposed spatial limits in a homogeneous system would have introduced difficult problems for prebiotic chemistry. Constructing boundaries, transposing things across them and modifying them when necessary are tasks accomplished today by the most refined cellular processes" (Eigen et al. 1981:91–92). In contrast, most metabolic theories, beginning with Oparin's pioneering coacervates and continuing with Fox's microspheres, assume the early emergence of a compartment. Such an enclosed volume, a protocell, is supposed to have served as an appropriate milieu for metabolism to develop. Metabolism is then postulated to have brought about the physical growth of the compartment and its division into two daughter cells upon reaching a certain limiting size. This process, seen by the metabolists as an inaccurate reproduction of the parent protocell, is supposed to have provided the basis for natural selection among daughter compartments and for their subsequent evolution.

Although they postulate an early partition, the theories of Oparin, Fox, and Kauffman view the primitive membrane as a passive device and focus on the internal content of the protocell, whose growth and division are regarded as being brought about by processes taking place within it. A different emphasis altogether is expressed by origin-of-life researchers who point to the crucial roles of the "shell" rather than the "core" in the emergence of life (Bachmann et al. 1992:59). Their theoretical arguments and experimental data suggest that lipid membranes assembled spontaneously on the primordial Earth to form vesicles, enclosing the aqueous soup medium within them. These membranes actively participated in the emergence of life by accommodating a source of energy for emerging life, providing a surface to support emerging metabolism, and being involved in self-replication (Morowitz et al. 1988; Morowitz 1992; Deamer 1997; Norris and Raine 1998). Membrane advocates consider the emergence of an entity separated from its environment by a membrane as the defining stage in the transition from nonlife to life.

The various hypotheses of the emergence of a membrane-bound protocell resemble other metabolic theories in considering information embodied in genetic polymers a late development in the emergence of life, a consequence of rather than a prerequisite for prebiotic evolution (Morowitz 1992:154). Unlike the other metabolists, however, membrane theorists and experimentalists focus on the early prebiotic synthesis and crucial role of lipids rather than amino acids and proteins. They also suggest that membranes participated in transforming light energy into chemical, electrical, or osmotic energy, thus driving the growth of protocells and the performance of metabolic processes inside them (Morowitz 1992:152).

To illuminate the lipid-protocell conception of the emergence of life, I will briefly introduce the scenario described by Harold Morowitz in *Beginnings of Cellular Life* (1992), which is based on major postulates and claims by several researchers, notably David W. Deamer of the University of California at Santa Cruz. The scenario starts with the assumption that the prebiotic environment contained hydrocarbons, some of which consisted of long chains of carbon-hydrogen groups. These molecules, either produced on the Earth or brought by meteorites (Deamer and Pashley 1989), accumulated on the surface of the ocean. There they interacted with minerals to produce compounds called amphiphiles (phospholipids are an example), which are characterized by their dual chemical nature: one end of the molecule is polar and hydrophilic, and the other is non-polar and hydrophobic. The amphiphiles condensed into various structures, among them mono- and bi-layers or sheets. The most thermodynamically stable state of an amphiphilic bilayer in an aqueous environment is reached when it closes to form a vesicle. In such a structure, the polar heads of the two layers in the amphiphilic membrane are pointing outward, toward the external aqueous solution, and inward, toward the aqueous interior. The nonpolar tails of the two layers are pointed toward the core of the membrane. This is the basic structure of extant biological membranes, which perform active transport of chemical compounds in and out of a cell through the participation of proteins attached to the lipid bilayer.

Morowitz contends that the formation of closed vesicles "represented the origin of supramolecular entities, three-phase systems consisting of a polar interior, a nonpolar membrane core, and a polar exterior—the environment" (Morowitz 1992:173). As experimentally demonstrated in model systems, Morowitz points out, the unfolding of origin-of-life processes depended on the properties of the amphiphilic vesicles. Nonpolar molecules, among them various chromophores (pigment systems) that absorb visible and ultraviolet light, tend to dissolve in the nonpolar lipid core of the membrane. Light energy is transformed by the chromophores into electrical potential energy, which can drive various chemical reactions, such as the synthesis of more amphiphilic molecules (Morowitz 1992:174–175). In extant cells, this reaction, and metabolism in general, are driven by phosphate-bond energy. The scenario assumes that this role was performed in the prebiotic environment by pyrophosphate. Whereas the original membrane vesicles are supposed to have formed from rare organic amphiphiles in the prebiotic soup, the energy produced once the vesicles existed ensured the synthesis of amphiphiles from common precursors in the environment. The next crucial stage in the development of the protocell was the incorporation of the newly synthesized amphiphiles into the existing membrane, a process that caused the vesicle to grow (Morowitz et al. 1988:284). Upon reaching a certain size range, the vesicle broke down to form smaller, more stable vesicles (Morowitz 1992:152). This step is considered by proponents of the vesicle scenario to represent replication of the protocell (Morowitz et al. 1988).

It is interesting to note that in a recent experiment, amphiphiles synthesized by the hydrolysis of an organic ester self-assembled to form micelles, a smaller, simpler version of vesicles bounded by a monolayer surface rather than a bilayer membrane (Bachmann et al. 1992). In the same reaction mixture, the micelles were then shown to catalyze the hydrolysis of more ester molecules to form additional amphiphiles, which self-assembled to form more micelles. Raising the question of whether this autocatalytic process is an appropriate prebiotic model, the authors acknowledge that micelles in themselves "are too simple, too small and dynamically too unstable to offer a plausible model for protocells." However, since under certain conditions micelles can be transformed into vesicles, the authors believe that their results are of significance for prebiotic studies: "Because of the simple mechanism underlying their spontaneous formation, aqueous micelles are plausible candidates for the first self-replicating bounded structures" (Bachmann et al. 1992:59).

As noted before, the major argument against any metabolic scenario that postpones the emergence of genetic polymers to a later stage has to do with the ability of a system lacking a self-replicating genetic molecule to engage in natural selection and to evolve. I have discussed Kauffman's and Wächtershäuser's answers to this challenge. The lipid-protocell scenario relies on the facts of vesicle growth and division, conceived as replication or reproduction. In order to describe a process of evolution, this scenario has also to account for a source of variations and for their persistence through protocell divisions. Morowitz suggests modes by which differences among protocells in a population persist with no need for genetic polymers. These "memory modes" are initiated by rare events (mutations) that lead to autocatalysis. If by being absorbed on the protocellular membrane a molecule enhances the transport of precursors of its own synthesis, such a memory mode becomes possible. Another possible mode is the initiation of an enhanced energy transduction (transformation) in the membrane, which helps the synthesis of the transducing molecule itself. Morowitz points out that if the rate of such autocatalytic processes is equal to or greater than the rate of protocell division, the memory of these events and processes will persist through repeated divisions. The retained activities could then serve as a basis for selection among protocells according to their ability to grow and replicate and hence as a basis for evolution of more complex structures and functions (Morowitz 1992:153–154).

Critics question one of the major assumptions of this scenario, the prebiotic synthesis of lipids that can assemble to form bilayer membranes. As indicated by Maynard Smith and Szathmáry, membrane formation requires fatty acids with linear hydrophobic chains made of more than ten carbon atoms, but under plausible prebiotic conditions branched-chain fatty acids rather than linear ones tend to form (Maynard Smith and Szathmáry 1995:100–101; Norris and Raine 1998:524). The fact that lipid-like material extracted from the Murchison meteorite was shown to self-assemble to form membranous vesicles does raise the possibility that membranogenic compounds were brought to the prebiotic Earth from an extraterrestrial source (Deamer and Pashley 1989). As already indicated, a mechanism forming long-chain amphiphiles on a pyrite surface under prebiotic conditions was suggested by Wächtershäuser. In Wächtershäuser's autotrophic scenario, the lipid, water-repellent layer creates a protective environment within which polypeptides and nucleic acids have a better chance to be synthesized. It is his belief that Morowitz's heterotrophic scenario, according to which membranes assemble and grow out of organic molecules in the primordial soup, is untenable. A lipid membrane, Wächtershäuser contends, is permeable for water but impermeable for ions and hydrophilic organic molecules (Wächtershäuser 1992:107). Wächtershäuser points out that this "self-suffocation by the impermeability of the membrane of heterotrophic nutrients" is "a major paradox for all versions of a heterotrophic cell formation in a prebiotic broth" (Wächtershäuser 1992:108). In Wächtershäuser's theory, on the other hand, amphiphiles are built on the pyrite surface from small inorganic nutrients, which can pass freely through a lipid membrane.

Another bone of contention is the permeability of ionic nutrients. Wächtershäuser claims that whereas these essential constituents to the development of a protocell are barred from entrance by a closed lipid membrane, they are provided in his scenario by the mineral surface, which during the gradual process of cellularization is still part of the metabolist (Wächtershäuser 1992:108). Supporters of the lipid-vesicle scenario retort by pointing to various mechanisms that might have overcome the barriers posed by primitive lipid membranes. Experiments show, for instance, that neutral forms of amino acids permeate membranes much faster than charged molecules (Chakrabarti and Deamer 1994). Model construction indicates that ions can be transferred across membranes through deep thinning defects, which make the membranes orders of magnitude more permeable (Pohorille and Wilson 1995:44).

Wächtershäuser challenges yet another aspect of the lipid-vesicle scenario, claiming that it does not meet the criterion of membrane-vesicle stability. Vesicles produced in a lipid solution are thermodynamically unstable. Theory and experiment indicate that the requirements for stable cellularization are many and complex. Wächtershäuser suspects that "the process of cellularization may well be biochemically unique and that it cannot be the stochastic 'self-organization' process in a lipid broth depicted by adherents of the hetero-origin theory" (Wächtershäuser 1992:109).

The various metabolic theories examined in this chapter, including the lipid-vesicle theory, postulate an origin of life not based from its very beginning on a self-replicating genetic molecule. Criticism of the vesicle scenario notwithstanding, the focus on lipid-like constituents in the prebiotic environment and a strong bioenergetic perspective on the emergence of life combine to make the metabolic conception a more comprehensive alternative to the genetic conception (Deamer 1997). In discussing the results of their experiments, Bachmann, Luisi, and Lang raise the possibility that their focus on "shell replication" will eventually be synthesized with the complementary study of "core replication," which focuses on template-based replicating structures (Bachmann et al. 1992:59). In the same vein, there are quite a few indications in various current studies that point to a growing realization of the need for a synthesis between different lines of origin-of-life research and even between traditionally opposing orientations. More will be said on this topic in the next chapter. Here, I would like to mention only two examples of a suggested synthesis within the metabolic camp. First, a theoretical study of the prebiotic assembly of amphiphiles to produce vesicles raises the possibility that Wächtershäuser's surface metabolists may have been responsible for the production of significant concentrations of the required amphiphiles, thus overcoming the difficulties of synthesizing long-chained fatty acids in high enough concentrations in the primordial soup (Norris and Raine 1998:524). And in a second example, a group of researchers at the Weizmann Institute has developed a computer model that provides a kinetic analysis of chemical sets manifesting mutual catalysis. The model is an elaboration of Kauffman's ideas of a system of weak catalysts achieving a catalytic closure upon reaching a certain level of molecular complexity. Kauffman discusses systems consisting of a set of peptides, or ribozymes, or both that could have functioned as autocatalytic, evolving entities on the primordial Earth. The Weizmann study examines a set of catalytic organic molecules, such as amphiphiles, enclosed within a membrane vesicle. It demonstrates the ability of such a vesicle to expand and self-replicate and the capability of some vesicles in a population to replicate faster than others (Segré et al. 1998). By its application to amphiphile assemblies, this model, originally inspired by Kauffman's theory of self-organization and mutual catalysis, may provide a quantitative basis for several versions of the membrane-vesicle scenario.

13

THE EMERGENCE OF LIFE—
NEITHER BY CHANCE
NOR BY DESIGN

The historical development of ideas about the origin of life, like other conceptual developments, has not followed a linear path. Interactions among empirical observations, philosophical presuppositions, and various cultural factors often resulted in confusion and setbacks rather than an increased understanding of the subject. Describing such a historical episode in chapter 5, I pointed out the seemingly paradoxical effect that the growth of biological knowledge at the beginning of this century had on the attitude of scientists toward the origin-of-life problem. Following the rise of biochemistry, genetics, and other scientific disciplines that contributed to the growing understanding of the enormous complexity of the cell, the protoplasmic theory, which attributed the basic characteristics of life to the cell content and disregarded the role of the nucleus and the cell wall, was gradually rejected. At the end of the nineteenth century, Huxley and Haeckel could still postulate a relatively simple process of abiogenesis of homogeneous lumps of protoplasm on the primordial Earth. This notion could no longer be entertained when life was conceived to be manifested only at the level of the cell, which was increasingly recognized to contain many enzymes and intricate internal structures. The situation drove many scientists at the beginning of the twentieth century to avoid the problem of the origin of life. As voiced by the geneticist Herman Muller, the general feeling at the time was that the subject was "too taboo" to touch (Muller 1966:494). Proponents of the "organismic movement" in biology, who, during the first decades of the century, investigated various physiological systems and focused on their organismic or integrated nature, adopted the same defeatist attitude toward the question of the emergence of life. The fundamental riddle was, What mechanisms could account for the emergence of such an intricate organization?

In *Chance and Necessity*, Jacques Monod wrote: "It might be thought that the discovery of the universal mechanisms basic to the essential properties of living beings would have helped solve the problem of life's origins. As it turns out, these discoveries, by almost entirely transforming the question, have shown it to be even more difficult than it formerly appeared" (Monod 1974:132). Monod was referring to the fact that, because of the achievements of molecular biology, the origin of life is now conceived in terms of the emergence of proteins and nucleic acids, whose interdependent relationship gives rise to the chicken-and-egg problem, regarded by Monod as a "veritable enigma." In Monod's view, to be discussed later, the original biological organization was a product of pure chance, and the *a priori* probability of its emergence was "virtually zero" (Monod 1974:135–136).

THE NATURE OF BIOLOGICAL ORGANIZATION

The nature of biological organization and the question of its emergence constitute the philosophical core of the origin-of-life problem. It is not by chance, then, that the basic philosophical conflict between scientific and creationist approaches to the origin of life focuses on these issues. In distinction to other natural objects, like a mountain, a river, or a planet, any living organism manifests a functional organization of its various components, which are dependent on each other and on the system as a whole. As demonstrated not only at the level of the whole organism but even more clearly at the molecular level, organisms are built of intricate systems that seem to be designed to perform certain functions. This unique nature of organisms was evident to any keen observer of the living world long before the establishment of modern biology. Throughout history, it was the focus of attention of naturalists and philosophers who wondered about "organic form" and its origin, each according to his philosophical conception and to the available scientific knowledge.

Aristotle, who in the fourth century B.C.E. characterized organisms as purposeful organized wholes, regarded the different organic forms—the principles of organization that determine the identity of the different species—as eternal (Aristotle 1952a; 1991), and thus he did not deal with the difficult question of the origin of these forms. The situation was very different in the eighteenth century, when Immanuel Kant, pondering the nature of biological organization, commented on the circular, self-reproducing character of a living system. The organism, he said, is "both cause and effect of itself" (Kant 1987 [1790]: 64:249), as manifested in the processes of reproduction, growth, and regeneration. While Aristotle assumed the Greek conception of the eternal universe, Kant's intellectual context obviously was different. In the eighteenth century the Christian dogma of the separate creation of the biological species and their static existence since creation was beginning to be questioned, and Kant was already theorizing about the origin of biological organization. He posed the basic philosophical question, still pertinent today, regarding the origin of life: Can we explain, in causal materialistic terms, the production of an organized whole from its parts, taking into account that in such a system parts and whole are reciprocally dependent? (Kant 1987 [1790]:77:293).

Kant's answer was that materialistic, mechanistic explanations could account only for the production of a haphazard aggregate from its individual parts. Because of the discursive nature of our reason, we can fathom the formation of an integrated whole only as a result of external design, an idea or plan that preceded the construction of the product (Kant 1987 [1790]:77:288–294). Thus we have no problem accounting for the human production of artifacts. However, in contrast to artifacts, organisms are not externally designed, and the different parts of an organism seem to function according to an internal plan of design (Kant 1987 [1790]:65:253). As a consequence, we cannot give a causal explanation of the purposeful nature of biological organization or of its origin. Never, Kant declared categorically, could another Newton arise who would explain in terms of natural laws the production of "a mere blade of grass" (Kant 1987 [1790]:75:282).

This last conclusion was based not only on Kant's conception of our understanding but also on his notion of the nature of matter, which he viewed as basically inert, by definition lifeless (Kant 1987 [1790]:73:276). According to this conception, material structures can form through random mechanical associations. However, the idea of physical principles of self-organization, which work mechanistically without any guiding plan and yet can produce an organized whole, seems like a contradiction in terms. This is why the very possibility of the emergence of life from inorganic matter seemed absurd to Kant (Kant 1987 [1790]:81:311). As will be seen shortly, the current creationist claim that life could not have emerged through natural processes is based on the same reasoning. Physical and chemical processes that could have been involved in the emergence of life are conceived as the "random shuffling of molecules" (Hoyle and Wickramasinghe 1981:3). Disregarding the Kantian (and Aristotelian) distinction between external and internal design, it is further claimed by creationists that the only reasonable explanation for the origin of life is intelligent design. Kant's critical philosophy, on the other hand, precluded such dogmatic assertions, which, according to Kant, illegitimately cross the line between science and theology (Kant 1987 [1790]:68). He advocated a limit self-imposed by the biologist regarding the question of origins. Biology could function as a science only by assuming an original organization as given and by pursuing scientific, mechanistic procedures from this starting point onward (Kant 1987 [1790]:81:311).

A revolution in the scientific understanding of the complexity, functionality, and obvious design of living organisms was brought about by the rise of the theory of evolution. Whereas Darwin did not discuss the origin of biological organization in his published work, the evolution of this organization by the mechanism of natural selection was a cornerstone of his theory. Natural selection, by favoring and selecting those organisms within a population endowed with the most adapted structures to perform the various biological functions, gradually led to the evolution of more complex and more adapted biological structures. This Darwinian mechanism provided a naturalistic explanation of the apparent design of biological organization, which gradually came to replace the one offered by Natural Theology, which had been forcefully presented by the Anglican priest William Paley in his *Natural Theology*, published in 1802. Paley focused on the intricate arrangement of organs and processes in the living organism, like the exquisite organization of the eye, as the strongest evidence of God's design (Paley 1970 [1802]). Fully acknowledging the facts of biological design—in fact, relying on examples given by Paley—

Darwin nevertheless advanced an alternative explanation based on natural, material processes.

THE "CHICKEN AND EGG" REVISITED

As this book demonstrates, researchers still grapple today with the question of the emergence of the original organic whole. Though the first biological systems must have been much simpler than the most primitive forms of life known to us, nevertheless, even the first stages of either replication or metabolism assumed by origin-of-life theories had to manifest an integrated, cyclic nature in order to work. We encountered this difficulty in discussing the question, How could nucleic acids emerge without proteins, or alternatively how could proteins emerge without nucleic acids? The presumed solution to this chicken-and-egg problem, the RNA world, was shown upon analysis to be bogged down by another version of the same vicious circle: a ribozyme acting as a replicase could, in principle, solve the nucleic acid–protein dilemma, but for a replicase to emerge there was need for evolution through natural selection, which could come into being only on the basis of some form of replication. The same predicament is also encountered by metabolic models. Stuart Kauffman describes the puzzle in the following way: "In order to function at all, a metabolism must minimally be a connected series of catalyzed transformations leading from food to needed products. Conversely, however, without the connected web to maintain the flow of energy and products, how could there have been a living entity to evolve connected metabolic pathways?" (Kauffman 1993:344).

The biochemist Michael Behe, a prominent figure among the "new creationists," who advocate the idea that the origin of life was due to an "intelligent agent," notes that biochemical research has revealed the presence in every living organism of "irreducibly complex systems." Such a system has a distinct function, the attainment of which depends on the interaction of the system's various components in such a way that if a single component is removed the system ceases to function. It cannot be produced, Behe claims, by the mechanism of natural selection operating on chance mutations. It cannot be formed by slight, successive modifications of a precursor system, "because any precursor to an irreducibly complex system that is missing a part is by definition nonfunctional"

(Behe 1996:39). In Behe's view, since the chances for the production of such a system "in one fell swoop" are so small as to amount to a miracle, the only viable explanation of both the origin of life and important stages in its evolution is intelligent design (Behe 1996:187–205). Contrary to Behe's contention, most evolutionary biologists describe the evolution of complex, adaptive systems as resulting from the working of natural selection with other evolutionary mechanisms that Behe chooses to ignore. Origin-of-life scientists agree with Behe's claim that complex biological systems—in fact even functional biological polymers of specific sequence—could not have materialized by a "happy accident" at one stroke. At the same time, however, origin-of-life research consists in looking for a naturalistic alternative to the idea of the creation of life by a designer.

The different models and scenarios discussed in this book represent various attempts to "reduce" the "irreducibly complex" first organized cycle by suggesting possible mechanisms for its emergence. Against the creationist conception that life and inanimate matter belong to two different categories and hence there is a need for an external agent to bridge the gap between them, scientific research into the origin of life maintains that living systems emerged from inanimate matter by physicochemical mechanisms. This seemingly self-evident unifying theme should be emphasized in light of the numerous conflicts among origin-of-life theories that have been discussed in the last chapters. I have pointed out various empirical disagreements pertaining to the nature of the early atmosphere, the source of organic compounds, the location on Earth where the first living systems emerged and the temperature of this environment, and the conventional or radical role played by minerals in the emergence of life. I have focused repeatedly on the main conceptual bone of contention that relates to the very definition of a living system: Is life basically a "replication machine," and hence, did life start with the emergence of a self-replicating molecule, and only later, through mutation and natural selection, did "all the rest" develop? Or should the first living system be characterized as an integrated cycle of weak enzymes that sustained itself through the exchange of matter and energy with the environment under far-from-equilibrium conditions, with the later appearance of genetic material being only the consequence of this primordial metabolism?

Both division and common ground among the different theories can be brought to a focus, as a way of summary, with the aid of a powerful metaphor suggested by Graham Cairns-Smith.

THE ARCH AND THE SCAFFOLDING

Advancing his "mineral-genes theory," Cairns-Smith emphasized the intricate circular nature of biological organization, describing the organism as a system in whose core "everything depends on everything" (Cairns-Smith 1985:39). How can a collaboration among components evolve, he asked, when the whole system will work only when all the components are there and working? (Cairns-Smith 1985:58). Cairns-Smith believes that such complex, interacting biochemical machinery, which was already in place in our last common ancestor, could come into being only following evolution through natural selection. But for that, some form of replication was needed, and we seem to be encountering the chicken-and-egg problem again. Moreover, as already noted, Cairns-Smith regards the prebiotic synthesis of nucleotides and sequences made of them as being extremely improbable. To illuminate both the problem involved in the emergence of biological organization and its naturalistic solution, Cairns-Smith introduces a highly suggestive image—an arch of stones, representing biological organization—and asks whether an arch could arise gradually, one stone at a time. The architecture of the central biochemical pathways, he points out, is in fact far more complex than that of an arch because each "stone" is connected not only to two others but to many. It seems that no arch of stones could arise gradually, hanging in midair, but it could arise with the help of supporting scaffolding. Stones could be added gradually to construct a wall, and when the scaffolding was removed the arch would remain in place (Cairns-Smith 1985:59–60). Cairns-Smith's sums up by saying that "before the multitudinous components of present biochemistry could come to lean together they had to lean on something else" (Cairns-Smith 1985:61).

Since, according to Cairns-Smith, organic compounds were much too complex to be synthesized under prebiotic conditions, he believes, as we have seen, that the first "organisms" capable of undergoing replication and evolution were made of mineral crystals. Hence the idea of the scaffold in Cairns-Smith's scenario is more than a figurative device suggesting that a naturalistic alternative to the design notion does exist. Organisms based on organic chemistry developed at a later evolutionary stage on the preexisting clay scaffolding. After the "genetic takeover" by the organic genes, the scaffolding made by mineral genes disappeared without leaving a trace in extant organisms.

Though the language used by Cairns-Smith in describing his arch-and-scaffolding model includes teleological terms (the building of a stone wall *in order* to support an arch, the removal of the stone wall *so that* the arch structure remains), no intention or planning ahead are implied. Intricately organized biological systems arose on the primordial Earth, and the mode of their emergence is what is being dealt with. Whereas creationists deny a naturalistic process of emergence, science assumes such processes and seeks to translate this general assumption into specific hypotheses. "The building of a stone wall in order to" stands for prebiotic mechanisms on the basis of which "an arch" could have emerged. Indeed, I find that the images of arch and scaffolding provide powerful conceptual tools for the analysis of the various origin-of-life theories, and I would like to summarize some of the major points made in the last chapters and to reemphasize some of my claims by using these images.

Is this appropriation of Cairns-Smith's metaphors for a wider, and in fact different, use than the one conceived by him for his mineral model legitimate? A scaffold is defined as a temporary, external supporting device, removed once construction is finished. Cairns-Smith's point was just that: rejecting the common notion in the origin-of-life field that gradual processes of organic chemical evolution gave rise to more complex living systems, he suggested an inorganic, mineral start that later disappeared. Since most origin-of-life researchers, unlike Cairns-Smith, believe in the continuity of organic chemistry and biochemistry from the early stages of emergence to present organisms, it seems that his metaphor would not apply to their hypotheses. And yet any origin-of-life theory must wrestle with the daunting image of the arch. Any such theory, relying on possible prebiotic mechanisms, suggests various scaffolding stages, chemical processes that could have led to a functioning, self-sustaining arch-like organization. The need to think in terms of scaffolding stages stems from the fact that prior to the emergence of the common ancestor, which no doubt was already an "arch" and which resulted from a long process of evolution, many more-primitive "arch-like" structures probably arose, serving as scaffolding for later stages. The arch-and-scaffold model can be conceived as a common, unifying theme of all origin-of-life theories. At the same time, the distinction between genetic and metabolic scaffolding is clear: these different approaches aim at the same organized system, which already contained interacting mechanisms of metabolism and self-replication, by suggesting the use of different "stone walls."

Metabolic Scaffolding

Revisiting Stuart Kauffman's model as a representative of the metabolic approach, we are reminded of Kauffman's focus on the nature of the living system as a complex whole and on his claim that his model faces the challenge of explaining how such a complex whole could have emerged by natural means through the dynamics of integrated metabolic cycles (Kauffman 1993:285–288). Thus it may be contended that this model, by postulating spontaneous self-organization as a physical principle manifested in certain systems, dispenses altogether with the need for a scaffold, producing, so to speak, a "spontaneous arch." Nevertheless, Kauffman's autocatalytic network of catalysts is also a scaffold, an "arch-like scaffold," one could say, but still only a stepping-stone on the basis of which a biological system comprising processes of both genetic self-replication and metabolism could have evolved. Such a complex system could not have evolved without the participation of natural selection, and Kauffman's autocatalytic networks can be seen as a scaffold on which natural selection could have begun operating. In contrast to the traditional Darwinian view that biological organization could result solely from the action of natural selection on random variations, Kauffman's model suggests that certain patterns of order had to emerge before natural selection could exert its molding action. Against the conception, epitomized by Jacques Monod, of evolution as "a random chance caught on the wing" (Monod 1974:96), Kauffman regards the physical principle of spontaneous self-organization as responsible for channeling evolution along certain paths (Kauffman 1993:22–26).

The notion of the scaffold as a system within which processes of chemical selection set the stage for the action of natural selection is very prominent in Wächtershäuser's pyrite-pulled chemoautotrophic model. Processes of natural selection in this system depended on the emergence of pyrite-bonded autocatalytic metabolic cycles. These cycles, however, could not have developed without extensive processes of chemical selection on the pyrite surface, which were due to the differential propensity for detachment of the various organic constituents formed on the surface (Wächtershäuser 1992:96). Though natural selection among variants of the autocatalytic cycles was responsible for the evolution of more complex biochemical cycles on the pyrite surface, chemical selection had a major role in the life of the surface metabolist. In fact, the selection of constituents by detachment remained the dominant selection factor until the emergence of an archike system. Only at this stage, following cellularization and the development of an integrated system of genes and enzymes, did natural selection among cellular units become predominant (Wächtershäuser 1992: 115). Like Cairns-Smith's clay scaffold, the original pyrite core of Wächtershäuser's surface metabolist is supposed to have been discarded once its organic constituents had gained a certain level of evolutionary complexity. However, prior to this point, the surface metabolist is postulated to have functioned as a highly ordered scaffolding whose physical growth and chemical development led to the emergence of an arch. Analyzing the geochemical setting on the primordial Earth, as well as thermodynamic and kinetic considerations and the principles of surface chemistry, Wächtershäuser deduces in great detail all the further developments of the surface metabolist toward complexity and cellularization.

Both Kauffman's and Wächtershäuser's models, despite their differences, belong to the category of "gene-second" theories, which assume the emergence of genetic self-replication as a relatively late event in the origin of life. In both models, autocatalytic cycles play a crucial role in the growth and evolution of the emerging biological entities. On the basis of the dynamics of autocatalysis, Kauffman and Wächtershäuser suggest mechanisms for evolution through natural selection without a replicating and mutating genome. In this they differ from earlier gene-second theories, Oparin's and Fox's, which focused on prebiotic processes leading to the establishment of peptide catalysts and metabolic cycles and did not pay serious attention to the mechanism of evolution or the role of natural selection. An interesting comparison can be made to early gene-first theories, such as those guiding the early experimental studies in the laboratories of Eigen and Orgel. These experiments focused on the evolution of optimal self-replication through natural selection, assuming the non-problematic synthesis of ribonucleotides and RNA templates and the non-enzymatic self-replication of RNA.

These early models did not pay serious attention to the constraining effects of prebiotic chemistry on the emergence of RNA and self-replication (Orgel 1973; Eigen et al. 1981). Current gene-first theories, the RNA- and pre-RNA-world scenarios, however, deal in detail with the need for a chemical scaffolding on which a self-replicating ribozyme and the RNA world could have emerged.

Genetic Scaffolding

The concept of the RNA world, suggested in the late 1960s prior to the existence of any direct evidence in its support, was itself a theoretical scaffold constructed to resolve the chicken-and-egg problem. Naturally occurring ribozymes, which were discovered later, and the additional ribozymes isolated in molecular-evolution experiments serve as evidence that strengthens the theoretical construct. The realization that an even more primitive scaffolding was probably needed for the emergence of a self-replicating RNA system led to the search for simpler, RNA-like genetic systems. Other models along similar lines suggested preliminary self-replicating systems completely different in nature from RNA. Cairns-Smith's mineral-genes theory belongs to this category. In more recent developments, already commented upon, some researchers insist on the evolutionary advantages of an early appearance of short RNA oligonucleotides for the origin of life. Consequently, these scientists suggest an elaborate system of chemical scaffolding to overcome the difficulties involved in the prebiotic synthesis of ribonucleotides and RNA sequences and in their replication. The scenarios of Kenneth James and Andrew Ellington (1995) and of Anastassia Kanavarioti (1994), for instance, serve as good examples of recent theories expounding the role of scaffolding in the origin of life.

The reasoning behind these scenarios is based on the logic of evolutionary continuity and selection. First, it is argued that since RNA sequences are superior in their ability to self-replicate compared to any other suggested substitutes and since they were definitely selected at some historical point to serve as the genetic molecules, it is reasonable to look as early as possible for their emergence as replicators (James and Ellington 1995:520). It is further assumed that the synthesis of nucleotides or of similar analogues, and of short sequences made of them, was probably catalyzed by minerals or peptides (Kanavarioti 1994:489–490). Once these sequences were formed, they could start to replicate using mechanisms that were probably different from the mechanism employed in extant organisms. Among early replication mechanisms, template-directed ligation is thought to have played an important role. The logic of selection comes into play when it is contended that among the varied population of sequences produced by the replication mechanisms, "pure," or homogeneous, sequences in terms of chirality and chemical bonds had many advantages over heterogeneous sequences. They could form, for instance, the most stable double helices and therefore had better chances of surviving and continuing ro replicate. On the basis of various experiments, it has also been concluded that RNA sequences might have been selected out of a heterogeneous pool because of their higher efficiency in binding substrates and catalyzing different reactions (James and Ellington 1995:520, 524; Kanavarioti 1994:488).

Converging Grounds

A common element in all the genetic versions of scaffolding is the reliance on catalysis, both in the synthesis of building blocks and polymers at any of the stages leading to an RNA world and in the different template-directed reactions. The role of catalysts in prebiotic chemistry was probably to enhance a limited number of reactions out of the wide spectrum of theoretically possible ones and to enable the synthesis of polymers of a more specific structure. The results of many experiments indicate catalytic activity of several metal ions, mineral surfaces, and peptides. The extensive role allotted by many genetically inclined researchers to mineral surfaces in the enhancement of the formation of peptides and oligonucleotides and as catalysts in various reactions has already been discussed. Here, in a similar fashion to Cairns-Smith's and Wächtershäuser's models, we are dealing with the physical aspects of prebiotic scaffolding. The catalytic role of protein-like polymers was already recognized in the early 1970s, following the work of Sidney Fox and others who demonstrated that amino acids and peptides or polymers were much easier to synthesize under prebiotic conditions than nucleic acids and that peptides may act as weak enzymes. I have pointed out that the emergence and functioning of Eigen's quasi-species and hypercycle were dependent on primitive protein-like catalysts, such as Fox's proteinoids. RNA-world scaffolding relies heavily on catalytic peptides. The role of protein-like enzymes in gene-first models and the similar part played by ribozymes

in several metabolic models (Kauffman 1993) point to the at-
tenuation of the strict division between genetic and metabolic
camps.

A telling example of an emerging overlap between traditionally separate points of view is the recent study by David Lee, Reza
Ghadiri, and their colleagues at the Scripps Research Institute
in which self-replication of a specific peptide was achieved via
template-directed ligation (Lee et al. 1996; Lee et al. 1997). In distinction to nucleic acids, whose replication is based on molecular
interactions between the complementary bases, here the peptide
serving as a template acted autocatalytically in replicating its own
synthesis from two peptide fragments, each identical in sequence
to its recognition site on the template. The researchers show that
peptide replication depends on both the structure and the sequence
of the polypeptide sequence (Lee et al. 1996). The study also produced a peptide hypercyclic system, "in which two otherwise competitive self-replicating peptides symbiotically catalyze each other's
production" (Lee et al. 1997:592). The researchers point out that
their results are of relevance to the origin-of-life theories of both
Eigen and Kauffman, possibly indicating that peptides could have
served in the origin of life both as carriers of information and as
enzymes (Lee et al. 1997:593).

Another indication of a possible joining of forces between traditionally theoretical rivals is to be found in several suggestions by
people who work within the RNA-world framework of a non-genetic
scaffolding leading to the emergence of a self-replicating genetic
system. Gerald Joyce, who believes that the RNA world must have
been preceded by elaborate chemical processes, discusses the possibility of alternative genetic systems as RNA precursors. At the
same time, however, he also considers the possibility of a period
of chemical evolution in which complex proteinic structures, which
were formed by non-genetic processes, could have altered the
chemistry of the environment in directions favorable to the future
emergence of a self-replicating system (Joyce 1989:222). Albert
Eschenmoser raises the possibility that autocatalytic metabolic
cycles, such as those conceived by Kauffman, Wächtershäuser, and
Christian de Duve (whose model will be shortly discussed), served
as an early stage, leading to the emergence of a genetic system. Even
if such autocatalytic systems, says Eschenmoser, did nothing but
grow, or accumulate a certain type of organic material in a given

locale, it would change the chemical environment and thereby influence the chances for the emergence of a genetic system (Eschenmoser 1994:393). Eschenmoser is talking about prebiotic processes of self-organization in a far-from-equilibrium environment, made possible by the availability of building materials and energy. Under these conditions, self-organization is favored thermodynamically and can be seen as necessary: through the dissipation of energy by the self-organized systems, the environment tends to return to equilibrium. These processes are necessary, says Eschenmoser, not despite but because of the Second Law of Thermodynamics (Eschenmoser 1994:389–390). The construction of a chemical scaffold—mineral, RNA-like, or metabolic—though involving many random events, is thus seen as "narrowly defined by the strictures of prebiotic chemistry," to use the language of James and Ellington (James and Ellington 1995:528).

The more specific and detailed the constraints that might have been imposed by the physicochemical setting on the early Earth, the fewer "degrees of freedom" are left for theories of the emergence of life. This is clearly demonstrated by the recently developed hydrothermal-systems scenario. In the strong focus on a specific geochemical setting for the emergence of the first living systems, the differences between genetic-oriented and metabolic-oriented theories seem to narrow. The theory of an iron-sulfide pulled-chemo-autotrophic origin developed by Günter Wächtershäuser describes the first organism as a surface metabolist and deals mainly with the emergence of metabolism. In the theory of crystal genes as originally proposed by Cairns-Smith, on the other hand, the first organisms were regarded as "naked genes" made of clay minerals, able to evolve through processes of replication, mutation, and natural selection (Cairns-Smith 1985:66). However, turning to the emergence of life in hydrothermal systems, Cairns-Smith, together with Allan Hall and Michael Russell, speculates that vesicles made of iron sulfide could have functioned as primitive cell-like structures. The semipermeable membranes confining these structures could have generated thermodynamic gradients and may have later been replaced by membranes made of organic molecules (Russell et al. 1988; Cairns-Smith et al. 1992:167–168). Within the hydrothermal-systems framework, Cairns-Smith's former genetic view is extended to include not only mineral genes, but also mineral catalysts and membranes. The original idea of genetic takeover is also extended to include the takeover, at a later evolutionary stage, of the role of catalysis by peptides and the role of membranes by lipids (Holm et al. 1992:188). It is plausible to speculate that the weight given in the future to metabolic versus genetic models will be largely decided by experimental work. For instance, experiments on the synthesis and stability of organic molecules and polymers, such as peptides and oligonucleotides, under high temperatures and pressures characteristic of hydrothermal systems are likely to constrain even more the origin-of-life models considered relevant and significant.

The growing role of empirical knowledge and specific considerations pertaining to actual primordial conditions should not mislead us as to the crucial part played by philosophical assumptions and theoretical postulates in guiding investigation. Yet examination of current developments in the field seems to indicate that traditional dividing lines between the genetic and metabolic positions are being redrawn. Judging by the present situation, future origin-of-life scenarios will be shaped more by demands of compatibility with prebiotic constraints than by preestablished philosophical commitments. The driving force behind this development is, I believe, not only the greater weight put on experiments but also the realization of the sheer enormity of the obstacles and the need to marshal every theoretical device to assure a breakthrough.

THE "HAPPY-ACCIDENT" APPROACH

Philosophically, the concept of the scaffold, as originally suggested by Cairns-Smith and as embodied in various origin-of-life theories, is an alternative to the idea of intelligent design. It is also an alternative to the idea that life emerged as a result of a chance event, or "happy accident." Historically, the notion that life emerged following a highly improbable, fortuitous "first event" was associated with the early gene-first theories. It was estimated that such an improbable event could have happened during the extremely long period of time available for the emergence of life. Furthermore, all that was needed, as Leonard Troland put it at the beginning of the century, was one successful collision of molecules. A single copy of the first gene, starting to replicate, would soon produce many more copies (Troland 1914:104–105). A minor statistical event would thus turn into a solid prebiotic fact. This "chance conception" was seriously entertained as long as the nature of genetic material was unknown. Later, in the 1960s and 1970s, during the era of optimism in the study of the origin of life, it was still upheld, though with greater difficulty. Nowadays, when the primordial geochemical conditions and the details of prebiotic chemistry are more realistically evaluated and the complexity of even very primitive biological systems is better understood, the chance conception is rejected by the majority of origin-of-life researchers. The improbabilities involved in a chance scenario are just too high for serious consideration (Fry 1995).

It is interesting, however, to note that the idea of a chance-like, highly improbable origin of life is still entertained by several prominent scientists. I have already commented on Francis Crick's theory of directed panspermia, which sprang out of his realization of the enormous difficulties involved in the emergence of life on Earth. Though proclaiming our ignorance as to the actual probability of this primordial process, Crick tends to regard the origin of life as a "happy accident" (Crick 1981:39), which was "almost a miracle" (Crick 1981:88). The renowned Harvard evolutionist Ernst Mayr asserted that "a full realization of the near impossibility of an origin of life brings home the point how improbable this event was" (Mayr 1982:584). The emergence of the interdependent linkage between nucleic acids and proteins was a source of perplexity for Mayr (Mayr 1982:583), and probably explains his notion of a unique, nearly impossible origin of life. In a more recent publication Mayr seems to have adopted a different approach to the issue, noting that "molecules that are necessary for the origin of life . . . have been identified in cosmic dust . . . so that it would seem quite conceivable that life could originate elsewhere in the universe. Some of the modern scenarios of the origin of life start out with even simpler molecules, which makes an independent origin of life even more probable" (Mayr 1995:152). Instead of pointing out the improbability of an origin of life, Mayr now focuses on the improbability involved in the evolution of human (and human-like) intelligence.

Though the views of Jacques Monod on the subject were published in 1970 (Monod 1974), when origin-of-life research was taking its first steps, the philosophical significance of his conception still deserves a serious comment. Monod's position was motivated by his rejection of any ideology or worldview depicting historical processes as teleological, that is, leading to progress and the appearance of man as the pinnacle of creation (Monod 1974:39–43). In his attempt to suggest an alternative, Monod described the starting point of biological evolution, the emergence of life, as a unique accident. We living creatures, he claimed, are here not because of any physical process leading to life's emergence but only because "our number came up in the Monte Carlo game" (Monod 1974: 137). Monod was not daunted by the chance formation of proteins and nucleic acids. He was struck by the "veritable enigma" of the emergence of a linked system of proteins and nucleic acids embodied in the genetic code. It is quite possible, he said, that the *a priori* probability of the random formation of such a system was "virtually zero." Monod acknowledged the unfortunate bearing of this claim on the ability of science to investigate the origin of life (Monod 1974:136). It is clear that a unique, highly improbable event can hardly be reconstructed and studied in the laboratory.

Monod based the "Monte Carlo conclusion" not only on his ideological position but also on the evaluation of the sequence and structure of cellular proteins. On the one hand, he emphasized the purposeful ("teleonomic," as he said) and non-random nature of extant organisms and the macromolecules they contain (Monod 1974:23–31). He pointed out that proteins are wonderfully elaborate machines whose function depends on their structure, which in turn depends on their basic sequence. The sequence of every protein is reproduced accurately in each cell, over and over again, in order to guarantee the invariance of its structure (Monod 1974:82–93). However, the sequence of every protein or polypeptide chain, in itself, shows no regularity, special feature, or restrictive characteristic and follows no rule but "the law of chance." The conclusion must be reached, Monod claimed, that this random sequence of a protein chain discloses "the pure chance of its origin," whereby no physical or chemical restrictions or constraints contributed to the original synthesis of proteins. Moreover, the origin and development of the whole biosphere is reflected in the origin and development of a functional protein: it all started with "the play of blind combinations" and grew into a purposeful, functional machine due to the operation of natural selection (Monod 1974:94–96). Monod's analysis shows that he acknowledged only two possible explanations for the emergence of life: either chance or purposeful divine creation. It is true, he said, that biological evolution, through natural selection, resulted in the teleonomic nature of biological systems, instructed by the genetic information encoded in the "genetic program." However, when discussing the stages prior to the establishment of biological systems, Monod did not consider the possibility of material mechanisms different from

chance associations between molecules. He did not foresee the likelihood of processes of self-organization suggested today by different origin-of-life theories.

Though acknowledging the role of contingent events in the
processes that led to the emergence of life, origin-of-life researchers
reject Monod's hypothesis that an "arch-like" primitive biological
system arose by the throw of the dice (Eigen 1992:11). The probability of the random emergence of a single cell is comparable, as
was pointed out by the British astrophysicist Fred Hoyle, to the
probability of the assembly of a Boeing 747 by a tornado whirling
through a junk yard (Hoyle and Wickramasinghe 1981). Furthermore, the notion that a primitive polypeptide or polynucleotide
arose by chance—that all the alternative sequences were physically
equivalent and that no constraining factor or organizing principle
limited the scope of all theoretically possible sequences—is also totally unacceptable. To reiterate our previous analyses, the reciprocal of the number of all possible combinations gives the probability
of the appearance of a particular sequence. Calculations for both
polypeptides and polynucleotides prove that the number of possible variants is beyond our realistic grasp. In the case of a polymer that corresponds to a single gene containing a thousand
nucleotides, the number of possible random variants is 10^{602} (Eigen
1992:10). The chance construction of the smallest catalytically active protein, made of one hundred amino acids, involves more than
10^{130} variants (Küppers 1990:60). Based on these numbers, the
chance formation of a specific polymer, though theoretically possible, is beyond belief considering available time and matter. As I
have repeatedly pointed out, origin-of-life theories rely on various
organizing principles, including selection mechanisms and catalysis, that are supposed to have limited and constrained the wide
scope of prebiotic chemical possibilities, thus constructing the scaffolding out of which the living arch eventually emerged.

The philosophical spirit underlying this scientific project is well
represented by the words of J. D. Bernal: "The question, could life
have originated by a chance occurrence of atoms, clearly leads to
a negative answer. This answer, combined with the knowledge that
life is actually here, leads to the conclusion that some sequence
other than chance occurrences must have led to the appearance of
life" (Bernal 1965:53). Current theories of the origin of life differ
in their suggestions of "sequences other than chance occurrences"
that will account for life's emergence. The goal of finding such a

sequence is explicitly acknowledged in the origin-of-life theory suggested by the Belgian biochemist and Nobel laureate Christian de Duve.

CHRISTIAN DE DUVE: AN EXPLICIT DENIAL OF MIRACLES

De Duve's model, described in detail in his book *Blueprint for a Cell* (De Duve 1991), belongs to the category of gene-second or metabolism-first theories. De Duve suggests that emerging life went through four successive "worlds": primeval prebiotic, thioester, RNA, and DNA. His contribution to the conventional RNA-world script, the insertion of the thioester world, is essential, he says, "because I cannot accept the view of an RNA world arising through purely random chemistry" (De Duve 1991:112–113).

Thioesters are sulfur compounds active in extant metabolic pathways, and De Duve claims that these compounds could have played an important role in the development of a primordial metabolism, or protometabolism. During the primeval stage, he postulates, a reducing atmosphere containing, among other components, hydrogen, methane, and ammonia arose following the action of ultraviolet light on ferrous iron salts, abundant on the ancient Earth. This reducing atmosphere could then bring about a Stanley Miller–type synthesis of different organic molecules, among them amino acids, other carboxylic acids, and thiols (alcohol-type compounds containing sulfur). At the next stage, the thioester world, amino acids and other carboxylic acids condensed with thiols to produce thioesters. These in their turn could have assembled to form short polymers ("multimers"), mostly made of amino acids, with the help of energy derived from the thioester bond. It is important to note that this process could have occurred without the help of a catalyst. According to de Duve, among the multimers were probably a number of crude catalysts that could have catalyzed the synthesis of various organic substances. Most important, some of these weak catalysts could have brought about the production of additional thioesters and multimers. Thus like other metabolic scenarios, de Duve's relies on autocatalytic cycles in the evolution of emergent metabolic systems: the catalytic peptides formed in the thioester world could also have catalyzed their own formation and the formation of other protoenzymes. The developing protometabolic networks came gradually to exploit the full catalytic potential of the multimer population. This eventually led to the synthesis of the first nucleotides and the first oligonucleotides and to the rise of the RNA world (De Duve 1991:115). De Duve's detailed scenario takes into account the physical and chemical conditions considered to have prevailed on the primordial Earth. Since thioesters tend to form under hot, acidic conditions, their primordial synthesis might have occurred in hot springs on the ocean's floor. Although this geochemical setting and the role allotted to sulfur compounds are reminiscent of Wächtershäuser's theory, it should be pointed out that de Duve's is a heterotrophic soup theory.

De Duve rejects the possibility of an RNA world emerging from random chemical processes. On the other hand, it is definitely possible, he claims, to describe mechanisms that led to the random synthesis of primitive protein enzymes. According to him, the term "random event or process" has two distinct meanings: the first and more popular relates to statistical randomness, the second to a process not directed by genetic information. If we describe the formation of a polymer from several kinds of amino acids, supposing that each of these kinds has an equal chance of being incorporated into the product, we are talking about a statistical random synthesis. Each of the constituents is represented in the polymer—the final product—according to its relative abundance in the substrate mixture. In such a random synthesis there are no distinctions among the different kinds of amino acids, and hence also no constraints that influence their sequencing in the polymer. Thus statistical randomness means sheer chance: starting again and again from the same initial conditions, a different sequence made of the same amino acids might arise, and there is no way to predict which product will be formed. When de Duve speaks of a random event or process in the origin-of-life context, however, he is not referring to statistical randomness but to the second meaning: not being directed by genetic information (De Duve 1991:141). Biological processes that occur in extant living cells are directly or indirectly informed by the sequence of bases in the genetic material of the cell. This is how enzymes are synthesized and how their function is to a large extent determined. Under prebiotic conditions, on the other hand, random synthesis of peptides took place initially without guidance from RNA segments. This randomness, claims de Duve, was very different from chance.

As noted above, the number of chance-like possible combinations of the smallest active protein (a sequence of 104 amino acids), which includes the twenty "species" or kinds of known amino acids, reaches 10^{135} (De Duve 1991:137; Küppers 1990:60). Only a very small percentage of these combinations could have been tried out during the prebiotic period, not only on Earth but in the entire universe. The chance that a specific combination constituting an active protein will be synthesized through statistical random means borders on the miraculous. But on the basis of various calculations, de Duve raises the possibility that short peptides could have formed randomly on the early Earth, provided they were made of no more than eight kinds of amino acids (instead of the current twenty kinds) and their length did not exceed twenty residues (five residues might have been the actual length) (De Duve 1991:137–140). Not only Sidney Fox's experiments but many others, among them studies conducted at Tel Aviv University by the astronomer Akiva Bar-Nun and his colleagues, attest to the ability of very short peptides, sometimes a mixture of amino acids or even certain single amino acids, to act as weak catalysts in different chemical reactions (Bar-Nun et al. 1994; Kochavi et al. 1997). The number of possible combinations of peptides of such limited composition and length is of course much smaller than the astronomical number mentioned above for a small but regular protein.

According to de Duve, however, as a result of stringent physicochemical constraints, only a very small number of peptides had the chance to be formed out of this limited group. As Fox already noted, there are significant differences in bond energies among different kinds of amino acids, and thus the bonds formed between different kinds of amino acids vary in stability. Because of this molecular selection, only a small percentage of possible peptides could have formed in significant amounts. Depending on the character of the participating amino acids, some of the peptides gave rise to insoluble aggregates and thus raised the relative concentration of soluble peptides that demonstrated better enzymatic activity in water. Since different peptides have different abilities to fold, they formed structures that varied in stability and enzymatic activity. In addition, active peptides that combined with their substrates were protected from breaking up. As a result of the various selective factors, only very rare peptides could have survived, and these survivors were generally more apt in their catalytic behavior (De Duve 1991:140–145).

The situation described by de Duve leaves very little room for chance: the possible combinations of amino acids, which due to the various constraints were only a tiny part of all the statistically possible combinations, were materialized. In contrast to a chance-like process in which there is no way to foresee the outcome, here we have a deterministic situation in which the products were determined unequivocally by the initial conditions. Hence, though the synthesis of primitive peptide catalysts and the establishment of a thioester world resulted from random processes, not being guided by genetic information, they were definitely not due to chance (De Duve 1991:213). This, however, could not have been the case with the emergence of the RNA world. De Duve's claim that an RNA world could not have arisen through random chemistry is now understood to mean that though random prebiotic processes were highly determined, that was not enough for RNA: the catalytic and energetic benefits of an established thioester world were needed to set the stage for the rise of RNA and self-replication (De Duve 1991:150–156).

De Duve's scaffolding, more explicitly than most other theories in the field, is an anti-chance construction. Considering the complexity of the biological arch and the many stages of the processes that had to be involved in its production, the alternatives, he says, are a natural, probable process or a miraculous creation (De Duve 1991:112). As I have tried to show, the various origin-of-life theories, metabolic and genetic alike, embody the same approach, if not to the same deterministic extent as de Duve's. Considering the chemical nature and organization of the macromolecules constituting living systems, and taking into account the constraints of prebiotic chemistry, the idea that life emerged by chance is equivalent to the claim that it was created by a miracle. Since the improbabilities are so high, this is still the case if the primordial terrestrial arena is enormously stretched and the whole material universe thrown into the picture (Eigen 1992:11). Thus, contrary to the ideas expressed by Monod, the two possible explanations of the origin of life are not "chance" and "purpose," the latter being designated as miracle, creation, or intelligent design. As will be shown in the following pages, the assumption that a natural emergence of life entails a highly improbable "chance association of molecules" often leads its holders to embrace an anti-natural, purposeful solution to the problem. Chance and telos, the two horns of a faulty dilemma, converge philosophically. The true conflict is between chance and purpose on the one hand, and a probable physicochemical process on the other.

"INTELLIGENT AGENT"—AN OPTION AMONG OTHER OPTIONS?

Sir Fred Hoyle and his colleague, the Sri Lankan scientist Chandra Wickramasinghe, are frequently mentioned in origin-of-life discussions as supporters of the idea of panspermia. In two books published in the late 1970s (Hoyle and Wickramasinghe 1979a; 1979b), they suggested that life originated in outer space and was brought to Earth—and is still being brought—by comets. They believed at that time that life emerged and evolved naturally in outer space long before the Earth was born, and has reached Earth "with the fundamental biochemical problems already solved" (Hoyle and Wickramasinghe 1981:8). According to Hoyle and Wickramasinghe, Earth has been showered with genes of cosmic origin, which have arrived here as pieces of genetic material, cells, or viruses. This cosmic genetic influx injected into Earth bits of information that were responsible first for the appearance of the earliest, already complex, living cells and then for all the major steps in biological evolution. A bizarre theory that saw the occurrence of widespread epidemics in different historical epochs as being caused by cosmic genes was also included in this general panspermic outlook.

Hoyle and Wickramasinghe resorted to the idea of panspermia following the realization that the first living systems, found in the oldest rocks on Earth, were already enormously complex. Furthermore, they believed that the natural processes that could have been involved in the origin of biological systems were "the random shuffling of simple organic molecules." Hence they concluded that the chance of these systems being naturally formed on Earth was "exceedingly minute" (Hoyle and Wickramasinghe 1981:3). Instead they relied on the wider cosmic arena to turn the highly improbable into a possibility.

Yet, as already indicated, once the idea of chance as the "natural mechanism" of life's emergence is brought in, even enlarging the physical arena does not help. As Hoyle and Wickramasinghe report in *Evolution from Space* (1981), they calculated the probability of producing an original set of enzymes by the "random shuffling of amino acids" to be "one part in $10^{40,000}$" (Hoyle and Wickramasinghe 1981:129). The conclusion drawn in this book, unlike their previous view, is that reliance on "a bigger and better organic soup" has to be forsaken. In light of the staggering numbers even a "universal soup" will not help. Natural, that is, chance-like, processes here or elsewhere, they now say, cannot explain the origin of life, and the only possible solution is to assume intelligent creation (Hoyle and Wickramasinghe 1981:150). As against the chance option, Hoyle and Wickramasinghe present the option of purpose—the intentional formation of life. In *Evolution from Space,* in *Cosmic Life-Force* (1988), and in the prologue to *Our Place in the Cosmos* (1993), looking to the whole universe is presented as offering the possibility of high intelligence within the universe "that is not God" and of many increasing levels of intelligence converging toward God as an ideal limit, never to be reached in practice (Hoyle and Wickramasinghe 1981:31). Hoyle and Wickramasinghe claim that their picture does not necessarily imply a purpose, but their view nevertheless is that there is a purpose: "The purpose is to generate life, not just on the Earth, but everywhere that it will take root, and to do so in an extremely elegant way" (Hoyle and Wickramasinghe 1981:32)

"Scientific Evidence of Intelligent Design"

Hoyle and Wickramasinghe's views on the origin of life can serve as an instructive introduction to some of the more sophisticated ideas currently advanced by the school of "new creationists." The premise that natural processes are incapable of producing a complex, organized biological system and an expressed commitment to an intelligent, purposeful agent are the two basic elements that, implicitly or explicitly, define this worldview. These elements form the common ground between Hoyle and the new creationists, even though most of the creationists would probably have nothing to do with Hoyle's "graded intelligence" or with the purpose he attributes to the intelligent agency, which is to seed the universe with life. The new creationists constitute a growing intellectual and political movement, mainly in the United States, which represents the latest development in the long history of creationism in this country. Their strategies are different from those of Biblical literalists, who view Genesis as a scientific and historical textbook and whose claims for "creation science" are rejected by both scientists and mainstream religion. Following several court decisions, notably a 1987 Supreme Court ruling that creation science is in fact religion, creationists are now conducting an increasingly successful campaign to introduce "scientific evidence against evolution" into the public school curriculum. As explained by one of the major figures in the new creationism, Phillip E. Johnson, a professor of law at the University of California at Berkeley, creation science interprets the Bible literally, promoting the notion of a young Earth, and special creation in six days. Creationists on the other hand, he says, may believe that the Earth is billions of years old and that simple forms of life evolved gradually to become more complex forms, including humans, but they also believe that "a supernatural Creator not only initiated this process but in some meaningful sense *controls* it in furtherance of a purpose" (Johnson 1993:4, n. 1).

The new creationists, among whom are a number of scientists, claim to derive the need for supernatural intervention from what they consider the absolute failure of science to provide a naturalistic explanation for the origin and evolution of life. Moreover, they claim that their belief in an intelligent designer is based on "scientific evidence" or "experience," insinuating that an intelligent cause is on a par with natural causes, being just another, though definitely better, option among others. Regarding the origin of life, the new creationists adopt the following tactics. First, they rightly characterize the unique complexity of molecular biological systems, rejecting the possibility of their emergence by chance. Second, skimming through some of the recent origin-of-life models without serious regard to their detailed content, they enumerate the many empirical and theoretical difficulties facing the field and conclude that a naturalistic mechanism by which life originated is impossible. The option of a gradual scaffolding construction, on which scientific research into the origin of life is based, is portrayed as an illusion. Citing the failure of scientists to reconstruct in the laboratory a possible prebiotic process resulting in a biological system, the creationists make a categorical jump, offering God as a solution (Bradley and Thaxton 1994:191–196; Behe 1996:166–173). To highlight the supposedly hopeless state in which origin-of-life research finds itself, they call attention to the differences dividing the various theories and to critical remarks by prominent scientists acknowledging the difficulties faced by the field. The study of prebiological evolution, the creationists contend, is today "about where Darwin left it" (Johnson 1993:103).

Though I have focused on the origin of life and have not dealt with questions pertaining to biological evolution, it is worthwhile to point out that the creationist tactic for dealing with the evolution of "irreducibly complex systems" is similar to the one used for discussing the origin of such systems. The Darwinian explanation of the development of complex organs and functions by natural selection of random mutations is declared completely useless after being incorrectly portrayed as based on chance. Other mechanisms suggested by evolutionary biologists in addition to the mechanism of selection to explain the development of complexity are ignored or dismissed (Behe 1996:16–18, 228–229; Johnson 1993:32–43). Johnson tells us that in order to emphasize the "really important point in Darwinism," he decided to avoid the term "evolution" and instead describe the Darwinian theory as "the blind watchmaker thesis," after the famous book by Richard Dawkins (Johnson 1993: 168). The metaphor of the blind watchmaker indeed captures the spirit of Darwinism, which sets out to explain biological design without any recourse to a designer. Since evolutionary mechanisms are responsible for the development of apparent design, evolution is blind in the sense of not having any purpose in view. However, though involving a strong element of contingency, evolution is not blind in the sense of having no direction and groping in the dark. The metaphor of the scaffold that served my origin-of-life discussion applies also to evolutionary developments. The evolution of complex structures and functions, though not intended, is not haphazard but highly constrained and channeled (Gould and Lewontin 1979:593–597; Endler 1986:236–242).

Discussing the origin of life, the scientists Walter Bradley and Charles Thaxton argue for the existence of scientific evidence of intelligent design, citing one of the fiercest critics of origin-of-life theories, Robert Shapiro of the Department of Chemistry at New York University. According to Shapiro, all current theories are bankrupt, and there is a need for a new paradigm in the search for a naturalistic explanation for the origin of life (Shapiro 1986). Agreeing with Shapiro's judgment, Bradley and Thaxton find it unfortunate that he and other scientists of a critical mind restrict their considerations to natural causes (Bradley and Thaxton 1994:196). This persistence of origin-of-life scientists in defending the notion that life originated by natural means despite all their failed theories, they say, is no more than "a philosophical commitment quite apart from experience." On the other hand, they counter, the inference of an intelligent agent is made on the basis of experience (Bradley and Thaxton 1994:177). The conclusion of intelligent design, says Behe, indeed flows naturally from the biochemical data, "combined with consideration of the way in which we reach conclusions of design every day" (Behe 1996:193). We are told by Bradley and Thaxton, Behe, and others that the inference of an intelligent agent is based on the "analogical method": we have learned by experience to associate a particular type of effect with a certain kind of cause, be it a natural or an intelligent cause. We have learned from experience, so the argument goes, to associate the existence of a designed system with a designer (Bradley and Thaxton 1994:198–199). We witness all around us evidence of the intelligent design of artifacts by humans. Moreover, "we do have experience in observing the intelligent design of components of life," since modern biochemistry routinely designs biochemical systems (Behe 1996:219). Bradley and Thaxton further contend that the SETI project, the radio-astronomy search for extraterrestrial intelligence, demonstrates that we are ready to infer the existence of an intelligent source from messages judged to contain specific information and that science accepts intelligent causes within its domain (Bradley and Thaxton 1994:199).

Various arguments were brought in the past to counter the use of analogy in promoting the idea of intelligent design. Among the best known were those presented by the eighteenth-century Scottish philosopher David Hume in his *Dialogues concerning Natural Religion*, published in 1779 after his death. Among other arguments, Hume indicated the fallacy in inferring from a limited human mind and its ability to produce artifacts the existence of an infinitely powerful divine entity. Furthermore, the analogy between an intelligent human mind and an intelligent supernatural designer, said Hume, cannot be tested empirically and thus cannot be verified (Hume 1966 [1779]:15–25, 37–39). Examination of the natural world, he maintained, which reveals many instances of waste, lack of adaptation, and evils, could easily lead to the inference of an incompetent designer (Hume 1966 [1779]: 61–70). The new creationists, however, dismissing arguments that are unfavorable to their cause, continue to advance their analogical procedure. Comparing human intelligence and an intelligent designer, they in fact assume what they are supposed to prove. Within the framework of the naturalistic worldview, which their argument intends to refute, human intelligence and intention, and consequently the design of artifacts by humans, are seen as natural, not supernatural. According to this worldview, humans as intelligent agents are the result of a long natural process of evolution. Extraterrestrial intelligence, if it exists, is also a natural product of similar evolutionary processes, be their details as different from our own evolution as they may. Inferring from these intelligent agents the existence of a supernatural agent simply begs the question.

In their defense, Bradley and Thaxton claim that the presumed intelligent agent may be either natural or supernatural and that the distinction between the two options belongs to the realm of metaphysics. However, they cannot have it both ways: if the intelligent agent is supernatural, it cannot be compared to humans; if the agent is natural, it must itself have come into existence through natural processes. Assuming that this agent is at least as complex as its alleged products, then according to the creationists' own logic we are back to the old impasse whereby the emergence of "irreducibly complex systems" defies any natural explanation. A similar argument was presented by Hume in his *Dialogues* (Hume 1966 [1799]:34) and also by Richard Dawkins in *The Blind Watchmaker* (Dawkins 1986:316). Thus this hypothetical intelligent agent, this "it" (or rather "he," or maybe "she") could not have originated naturally. The contention that the idea of intelligent design does not necessarily involve supernatural connotations is eventually exposed for what it is. Bradley and Thaxton declare, at the end of their essay, that the most reasonable speculation about the origin of the first DNA molecules is that "there was some form of intelligence around that time" and it is most reasonable to posit that "life came from a 'who' instead of a 'what'" (Bradley and Thaxton 1994:209). Behe also attempts to have his cake and eat it too, claiming that science can put the identity of the designer "on the back burner" and wishing philosophy and theology well if they would like to try to solve the problem in the meantime (Behe 1996:251).

Conflict between Two Worldviews

In *Darwin on Trial*, Phillip Johnson engages in a misguided and highly uninformed attack on the theory of evolution, claiming that its main tenets lack any empirical basis. The same sweeping assertions are reiterated in his more recent book, *Reason in the Balance* (Johnson 1995:11). Johnson's caricature treatment of Darwinism has been severely criticized by the Harvard paleontologist Stephen Jay Gould (Gould 1992) and by other biologists and philosophers of biology (Hull 1991; Ruse 1996). Johnson's attempt to discredit the empirical basis of the Darwinian theory is accompanied by a misrepresentation of the complementary relationship between fact and theory in science and between the empirical and philosophical aspects of science. Questioning, rhetorically, whether Darwinism is either empirical science or philosophy and claiming that the theory of evolution is not part of empirical science at all, he concludes that it is nothing but a deduction from naturalistic philosophy (Johnson 1993:158).

Johnson's contention that the case for evolution has not been proved reflects his distorted conception of the nature of scientific work. Ignoring the fallibility of science—the fact that scientists err but can put their conclusions to the test—he regards controversy among evolutionists and the repeated need to reevaluate hypotheses about evolutionary mechanisms and specific evolutionary developments as clear evidence of the collapse of the theory of evolution. Discussing the nature of science, the philosopher of science Philip Kitcher points out that the creationist argument against evolution is based on a popular but mistaken notion. Science, Kitcher contends, is not a body of demonstrated truths, and scientific activity does rely on belief where we cannot prove. Scientific theories, however, including the theory of evolution, in contrast to dogmas based on pure faith, are open to revision on the basis of new empirical evidence. "Even though our present evidence does not *prove* that evolutionary biology—or quantum physics, or plate tectonics, or any other theory—is true, evolutionary biologists will maintain that the present evidence is overwhelmingly in favor of their theory" (Kitcher 1982:34). Addressing other misconceptions of science entertained by many working scientists and exploited by the creationists in their anti-evolutionary campaign, Kitcher indicates the characteristics of a successful scientific theory. These include not only the testability of hypotheses under a variety of conditions, but also the strength of a successful theory in unifying a broad class of previously unsolved problems and the ability of such a theory to open up new lines of investigation. There is no doubt that the Darwinian theory fulfills all these criteria (Kitcher 1982:35–54).

The scientific community commonly rejects attempts to discredit the scientific status of the theory of evolution. However, the new creationists' contention that Darwinism is based on naturalistic philosophy draws divided responses. Some scientists view it as an unjustified accusation. Still adhering to the discredited positivistic view of science as an accumulation of objective facts free from theoretical and metaphysical presuppositions, they would regard any admission that the theory of evolution has its own metaphysical basis as a sell-out to the creationists (Shapiro 1993). Philosophers of science usually acknowledge that science in general, including evolutionary biology, is committed to methodological naturalism. The methodological naturalist, they say, adopts the attitude of "pushing the scientific method (including the commitment to unbroken law) as far as [he] can" (Ruse 1996:573). Unlike an ontological, or metaphysical, naturalist, the methodological naturalist is not, however, committed directly "to a picture of what exists in the world" but only indirectly "to what the methods of the natural sciences discover" (Pennock 1996:549). These philosophers of science indicate that scientific methodology must depend on pursuing natural processes, rejecting supernatural causes. As it is put by Robert Pennock, "Lawful regularity is at the very heart of the naturalistic worldview and to say that some power is supernatural is, by definition, to say that it can violate natural laws" (Pennock 1996:552).

Acknowledging the necessary reliance on methodological naturalism, many biologists and philosophers of science nevertheless argue that since science as a rule is not able to comment on metaphysical questions, "the science of Darwinism is fully compatible with conventional religious beliefs" (Gould 1992:119). As proof that there is a way to embrace both creation and evolution, religious evolutionists are cited, notably among them Theodosius Dobzhansky, one of the greatest evolutionists of our time and a believing Russian Orthodox. However, peaceful coexistence between science and religious beliefs depends on the fulfilment of certain conditions. As Pennock points out, scientists have to renounce metaphysical naturalism. Among the theological views compatible with methodological naturalism he mentions two options: a deistic God who created the world but, thereafter, does not intervene in the natural order and a God who intervenes only at a spiritual level (Pennock 1996: 550). Some metaphysical naturalists, says Pennock (Spinoza and Hobbes, for example), who attempt to "naturalize theology" by taking God to be not necessarily supernatural, still find room for such a being. However, both traditionalists among religious believers and many naturalists who take for granted the concept of a transcendent God and adopt a strong secular stand object to this position (Pennock 1996:549). Dobzhansky's ideas on the subject might be considered an example of naturalization of theology. Like many evolutionary theists, he too voiced the notion that "Creation is realized in this world by means of evolution" (Dobzhansky 1973).

Those within the scientific community who insist that no conflict exists between science and religion often paraphrase the ideas discussed above by simply claiming that the lack of conflict arises "from a lack of overlap between [the] respective domains of professional expertise—science in the empirical constitution of the universe, and religion in the search for proper ethical values and the spiritual meaning of our lives" (Gould 1997:18). These scientists consider the address of Pope John Paul II in October 1996 to the Pontifical Academy of Sciences in Rome on the issue of evolution as a clear vindication of their views. They point out that John Paul II not only reiterated the 1950 message of his predecessor Pius XII that there is no opposition between evolution and the doctrine of the faith, he went even further in discussing the progress of science on issues relevant to biological evolution during the decades since 1950. The work of different fields of science, the pope said, is "a significant argument in favor of this theory." On the basis of this convergent evidence it is now possible "to recognize that the theory of evolution is more than a hypothesis" (John Paul II 1997:382).

The scope and nature of this book will not allow me to engage in a lengthy discussion of the relationship between religion and science, which involves complex philosophical, historical and political issues. Nor will I be able to explore in depth the implications of the distinction between methodological and metaphysical naturalism. However, I contend that more than any other area of biological research (with the obvious exclusion of the fields devoted to the evolution of human consciousness and culture), the scientific study of the origin of life involves a philosophical commitment that goes beyond questions of methodology. Confronting the problem of the origin of life brings to the fore the dividing line between scientific and religious conceptions of nature—questions of purpose, intentional intervention, and control of the natural world. While it might be possible to remain neutral on these questions when dealing with some aspects of the evolutionary process, it is definitely not so regarding the important juncture of the emergence of life. The lesson drawn from our discussion of the issue of intelligent design has relevance beyond the confines of the creationist position. Though some advocates of intelligent design try to conceal the teleological nature of their hypothesis by not committing themselves as to the identity of the designer, I believe that Phillip Johnson's views inadvertently provide good service to the Darwinian cause by calling attention to the core of the conflict, which is the idea of intentional intervention in nature. Johnson emphasizes the radical dichotomy between the evolutionary and creationist worldviews: whereas creationism stands or falls on the idea of a purposeful supernatural creator, the cornerstone of science in general and of the science of evolution in particular is the assumption that no supernatural intervention is involved in the natural world (Johnson 1993:91, 125–134).

The difference between the explicit creationist position, old or new, and the views expressed by mainstream religious thinkers, including the pope's recent statement, seems to be very clear. Not surprisingly, creationists of all shades have attacked the pope for his support of evolution, charging that he has compromised God's word and ignored the intelligent-design argument against evolution (Scott 1997:402). And yet the attempts made by several members of the scientific community to draw a line between the pope and science on one side and creationists on the other ignore the fact that, though conceding that evolution is "more than a hypothesis," the pope delivered a message antagonistic to the basic spirit of the evolutionary theory. Pointing out that the theory of evolution, like any other scientific theory, is consistent with observed data but also relies on philosophical tenets, he made it clear that he is not ready to accept a theory of evolution based on a materialist interpretation, but only one that is spiritualist in its philosophy (John Paul II 1997:382). Focusing on the creation of man and reiterating the position of Pius XII, John Paul II indicated that if the human body took its origin from preexistent living matter, the soul was created immediately by God. He conceded to the theory of evolution the option of explaining material continuity as far as the evolution of the human body goes, but flatly rejected any attempt to explain on scientific terms the novelty and uniqueness involved in the evolutionary emergence of the human species (John Paul II 1997:383).

Yet it is the very nature of the scientific explanation of the process of evolution, including the emergence of *Homo sapiens*, to account for both continuity and novelty and to do so in materialist terms. Moreover, the pope was addressing a crucial aspect of natural reality, making decisive claims about the nature of evolution and not confining himself to the realm of "spirituality." This state of affairs obviously refutes the contention that science and religion do not overlap in their respective domains. As Richard Dawkins states, "A universe with a supernatural presence would be a fundamentally and qualitatively different kind of universe from one without. The difference is, inescapably, a scientific difference. Religions make existence claims, and this means scientific claims" (Dawkins 1997:399).

Belief in a supernatural presence has strong implications for the way the evolution of the human species and the origin of life are viewed. It might be argued that the latter subject can be studied by pursuing the scientific method, without taking a radical metaphysical stand. My claim is that origin-of-life scientists, purposefully or inadvertently, implicitly or explicitly, by the very act of conducting their research are making a philosophical decision and taking a metaphysical stand. They commit themselves to a picture of a natural world in which material processes of self-organization could have produced life, and by doing so they reject the contention of the new creationists that the "high information content" embodied in any living system, including the most primitive, could have originated only by an act of intelligence. Bradley and Thaxton, who claim that "life came from a 'who' instead of a 'what'" (Bradley and Thaxton 1994:209), allege that one can affirm natural causes as the source of the origin of life and at the same time entertain the idea that "some higher power or intellect (God?) . . . stands behind the scene, guiding the 'natural' processes" (Bradley and Thaxton 1994:176–177). The fact that quite a few scientists, especially in the United States, choose to adopt such a position does not resolve the conflict between the mutually exclusive natural and supernatural explanations embodied in it. These two metaphysical explanations have entirely different consequences regarding physical reality, and consequently conflicting implications regarding the nature of scientific research. I maintain that it is as a result of their philosophical commitment to naturalism that researchers logically insist on pursuing the study of the origin of life when no satisfying solution to the problem has yet been found. Within the framework of the naturalistic worldview, they find no reason to abandon research and yield to the "easy-way-out" of intelligent design suggested by the creationists.

In his book *Steps towards Life*, Manfred Eigen raises the question of whether the possibility of a natural origin of life can be disproved. In order to do that, he says, "one would have to be acquainted with all the historically possible conditions, and then to show that there is no catalytic mechanism that under *any* of these sets of conditions could have carried out the desired task. Such a proof is hardly conceivable, on the account of the enormous number of possible mechanisms and conditions that would have to be excluded" (Eigen 1992:37–38). One could paraphrase Eigen's argument by saying that only God could prove that life could not have emerged naturally. Eigen goes on to point out that the biochemical mechanisms in every living cell serve as a disproof of anti-natural claims. All the known reactions taking place in every organism are fully explainable in scientific terms. It is highly conceivable that all of these reactions were preceded by similar, though simpler, prebiotic mechanisms (Eigen 1992:38). One could contend that Eigen's elaboration on the natural origin of life and the possibility of its denial should be interpreted within a methodological-naturalistic framework. Thus the need to continue the search for the desired mechanism under the appropriate set of conditions and the notion of evolutionary continuity between life's first steps and extant organisms can be seen as part of the methodology of science. And yet the methodological principles of science are justified on the basis of a certain view of nature, a view that relies on the laws and mechanisms of nature and has no room for an intentional "outside job."

The Need for Historical Perspective

The tendency to ignore the crucial philosophical role of the naturalistic standpoint in the scientific study of the origin of life is demonstrated in the frequent confusion between the status of empirical and philosophical claims. The confusion is cultivated by creationists claiming to present scientific evidence for intelligent design. It is also evident in the views expressed by scientists who somehow expect the role of a creator in the origin of life to be determined by scientific research. Christian de Duve was quoted recently as saying that "until such time as biologists can demonstrate an entirely material origin for life, the divine will remain a contender." Hope that future scientific research will resolve the conflict is proved futile by the words of the NIH geneticist Francis Collins, who, following de Duve's words, was quoted as asking "Why couldn't God have used the mechanism of evolution to create?" (Easterbrook 1997:893). Scientists can get blue in the face attempting to demonstrate "an entirely material origin for life" but still for those who do not adopt the naturalistic position there will not be proof. We need to understand the different epistemological status of empirical statements, scientific theory, and philosophical commitment, and their interaction. The intricate relationship among these entities, especially in the study of the origin and evolution of life, can be truly appreciated only when the historical development of this relationship is examined.

Creationists like to point out that science's exclusive commitment to naturalistic procedures stems from its blind reliance on metaphysical naturalism (Johnson 1993:158–159). They accuse Darwinism of being just a dogmatic faith in naturalism. It is true that science can neither prove nor disprove God's existence and that knowledge gained by pursuing the naturalistic method is by definition limited to those segments of reality that are open to our investigation. Nevertheless, creationists ignore the fact that the naturalistic worldview, including positions that go beyond the empirical scope of our investigation, developed historically through an intricate relationship with empirical knowledge. Naturalistic evaluation of nature is not a newborn required to justify anew its right to existence. Scientists making observations and performing experiments within the framework of the evolutionary conception do not see it as their goal to prove or disprove its basic tenets. This is due not to blind faith but to the confidence of science in its conceptual framework, a confidence reflecting the fact that the Darwinian worldview was built gradually as a result of many factors, not least among them the accumulation of empirical observations.

As an increasing number of natural phenomena lent themselves to the naturalistic method, within the framework of technological and social changes, a growing body of knowledge was produced and a naturalistic worldview became established. The theory of evolution, contributing to and strengthening earlier trends in astronomy, geology, paleontology, and other natural sciences, gave rise to a new scientific and philosophical conception that made it possible to examine nature from a naturalistic standpoint not possible before. The claims made on behalf of the naturalistic worldview had and still have a different status from those of empirical science: they cannot be tested experimentally, but they nevertheless provide science with the general framework for its work. The relationship between the grand evolutionary philosophical conception and empirical studies consists in an ongoing mutual influence and joint development. Thus the theory of evolution is indeed based on a naturalistic worldview that entails a metaphysical commitment, but one that developed on the basis of a sound empirical foundation and is being continuously strengthened by it.

No Case for Equal Time

The new creationists contend that since the theory of evolution has been empirically debunked and since Darwinism (naturalism) and creationism present two different philosophies, the two points of view deserve to be given equal time in high schools, universities, and society at large. I have already countered the claim for equal status by the historical argument: in distinction to creationism, the evolutionary worldview developed on the basis of cultural, political, *and* empirical factors. The theory of evolution not only unified the different branches of biology but in its wider connotations also provided a solid basis for the non-biological sciences and led to the development of new scientific fields. Such a contribution to the growth of human knowledge definitely cannot be attributed to creationism. Another crucial difference between naturalistic and creationist philosophical positions concerns the ability to expand our knowledge of the natural world. The fact that scientific activity involves putting empirical hypotheses to the test, while belief in a creator and designer of nature precludes such a procedure, seems self-evident. Most scientists thus consider scientific and religious projects as completely incommensurable (see, for instance, Eigen 1992:127). However, the new creationists claim not only to combat naturalistic philosophy and its deplorable consequences but also to advance scientific knowledge.

Scientists of good will who preach a "division of labor" between science and religion suggest that science takes care of factual reality while religion struggles with human morality (Gould 1992:120). But the intelligent-design proponents will have none of that. They present the creationist notion of intelligent design as a major breakthrough in modern science and as a "new theory" to boot! This preposterous claim, ignoring centuries of debate over the Argument from Design for the existence of God, is made, for instance, by Michael Behe. His claim to novelty in fact rests on dressing up the old design argument in molecular garb (Behe 1996:230). As more and more scientists grow curious about design, he predicts, the whole field investigating designed systems will be reinvigorated. However, when we examine Behe's suggestions as to which directions this future research could take, we come face-to-face with a "reinvigorated version" of ideas on the inevitable limits of biology put forward in the eighteenth century by Immanuel Kant. As noted at the beginning of this chapter, Kant realized that biological organization manifests design, and hard as he tried, preceding

Darwin and modern biology, he could not come up with any scientific causal explanation of the origin of this design. Not willing to admit theology into science, Kant concluded that biologists should treat the original organized system as given, exploring its further mechanistic development (Kant 1987 [1790]:81:311).

Toward the end of the twentieth century, Behe, unlike Kant, presents us with an odd mixture of science and mythology. He considers as possible the "simplest possible design scenario" that "nearly four billion years ago the designer made the first cell, already containing all of the irreducibly complex biochemical systems." One can however postulate, he continues, that the designs for systems that were to be used later, "like blood clotting," were present but not "turned on." In that case, research could be undertaken "to determine whether information for designed systems could lie dormant for long periods of time" (Behe 1996:228, 229, 231). The Kantian project as an inviting suggestion for scientists weary of naturalism is formulated even more clearly by Johnson, who says, "Why not consider the possibility that life is what it so evidently seems to be, the product of creative intelligence? Science would not come to an end, because the task would remain of deciphering the languages in which genetic information is communicated, and in general finding out how the whole system *works*" (Johnson 1993:112).

NEITHER BY CHANCE NOR BY DESIGN

Origin-of-life scientists will probably not consider these suggestions as serious and will continue their research, looking for "mechanisms other than chance" that could have brought about the self-organization of primitive biological systems under prevailing prebiotic conditions. Making a commitment to either a creationist or a scientific position on the question of the emergence of life is completely legitimate as long as the implications of the choice are clear. Commitment to the evolutionary philosophical worldview means that the only way to solve the origin-of-life problem is to continue scientific research. This is the case even though, as our discussion in the previous chapters has shown, no theory suggested so far has gained the general support of researchers in the field. Early stages of chemical evolution seem to be successfully established. There is evidence of the existence of primitive forms of life in the oldest rocks on Earth. Recent molecular studies have revealed a genetic record of the convergence of life toward a common ancestor, and the physical and chemical requirements of this postulated population of ancestral cells seem to match our knowledge of the conditions on the primordial Earth. Yet no single scenario has so far led to a reproduction in the laboratory of a possible synthesis of a living system. Researchers, however, do not regard this state of affairs as cause to abandon the scientific ship. The optimism characteristic of the field in the 1960s and 1970s was replaced by a more serious appreciation of the difficulties involved in any attempt at a solution. A better understanding of both the primordial geochemical setting and the molecular basis of life, while abolishing the option of a chance-like emergence, is giving rise to new creative approaches to the origin-of-life problem. Strong evidence of the confidence of science in a natural solution to this problem is provided by the conviction of the majority of scientists of the existence of extraterrestrial life. Based on a variety of recent findings, this conviction is increasingly being translated into an active search for locales in outer space where conditions that could have produced life prevail and where life itself has emerged.

14

"LIFE ON MARS?
SO WHAT?"

This provocative title was chosen by Stephen Jay Gould for his article in the *New York Times* on August 11, 1996. Only a few days had passed since posible evidence of past life on Mars found in a meteorite was announced at a press conference organized by NASA and broadcast on television to the whole world. Following the scientific reports, President Clinton described the discovery as one of the most important in the history of science. Gould, for his part, though "delighted beyond measure" with the news, wished to explain why he did not view the fossil evidence as a revolutionary scientific discovery and why in fact he was not surprised by the news. Pointing out that life arose on Earth "almost as soon as environmental conditions permitted," Gould went on to state "the standard expectation of paleontologists:" We can only infer from this rapidity, he said, "that it is not 'difficult' for life of bacterial grade to evolve on planets with appropriate conditions. The origin of life may be a virtually automatic consequence of carbon chemistry and the physics of self-organizing systems, given favorable conditions and the requisite inorganic constituents." Since, as Gould indicated, there is abundant geological evidence attesting to the presence of liquid water on Mars during its early history—a finding that also implies that Mars was once warmer than it is today—the conclusion is straightforward: "If bacterial life arose so quickly on Earth, and if Mars once possessed similarly favorable conditions, then we should also anticipate the evolution of some form of life at bacterial grade on Mars as well."

Gould regarded the evidence presented by the meteorite research team to be far from decisive and indicated that convincing evidence of a real Martian fossil would be of enormous importance to science. The main thrust of his argument, however, was the claim that life at bacterial grade is a universal phenomenon to be expected wherever and whenever appropriate conditions exist. I have already shown that this is the basic thinking underlying research into the origin of life. In a recent publication, Christopher McKay, an astrogeologist and expert on Mars at the NASA Ames Research Center in California, pointed out that the universal condition common to all origin-of-life theories is liquid water. Assuming that "the development of life is a fundamental and reproducible process and will occur given suitable chemical and physical conditions," the evidence of liquid water on early Mars makes it possible to discuss the origin of life on that planet within the framework of current terrestrial theories (McKay 1997:268). Furthermore, on the basis of data regarding geological and atmospheric conditions on ancient Mars, it is evident that the specific requirements for the origin of life on Earth as postulated by the different theories could also be met on Mars. According to McKay, "Both early Earth and early Mars had active volcanism and associated sulfurous hydrothermal regions, subsurface hydrological systems, small ephemeral ponds, large stable bodies of water, meteorite and cometary impacts, anoxic (oxygen-less) conditions and airborne particles and bubbles." His conclusion is that "if life arose on Earth then it could have arisen on Mars as well" (McKay 1997:269).

Not surprisingly, proponents of the intelligent-design hypothesis have denounced the naturalistic commitment to find signs of extraterrestrial life, which in their view might lead to the "illusion of knowledge." Bradley and Thaxton claim that it is this commitment that has led "to exaggerated expectations of finding organics, if not life itself, on Mars and accounts for the current optimism of those who are eager to return to Mars for another sample-gathering expedition"(Bradley and Thaxton 1994:198). I will describe current scientific studies examining the evidence of organic material and signs of life, past and present, on Mars and will show how far this sober and cautious enterprise is from "exaggerated expectations" or unchecked optimism. We witness here another demonstration of the philosophical conflict between pursuers of scientific research into the origin of life and the new creationists. Science, based on the naturalistic worldview, is not giving up the attempt to solve the origin-of-life problem, while the promoters of the design solution feel no need to return to Mars for more evidence, convinced as they are that an organized biological system, here or on Mars, could have emerged only by intelligent design.

Before continuing with the discussion of ALH84001, it is important to point out that most scientists that have been involved for the past few years in the investigation of this meteorite tend to think that available data in the rock cannot settle the question of past life on Mars. And yet the story of this meteorite, including the controversy surrounding its research, is worth telling, for reasons to be explained shortly.

FOSSILS IN A METEORITE FROM MARS? A CASE STUDY OF SCIENCE AT WORK

The meteorite discussed at the NASA press conference was found in 1984 in an ice field in the Allan Hills region of Antarctica by a research team organized by the Johnson Space Center in Houston, Texas. The rock, weighing 1.9 kilograms (4 pounds), was later labeled ALH84001. The ice fields of Antarctica proved an ideal site for spotting meteorites, most of which fall into oceans around the world, never to be retrieved. In Antarctica, falling meteorites are buried in the ice but can be pushed upward where the floating ice meets mountain slopes. As the ice is eroded by wind flows, the meteorites are exposed. Among the myriad rocks that continuously bombard the Earth, we know today of thirteen that have come from Mars, half of them found in Antarctica (Score 1997). The Martian origin of these rocks, including ALH84001, was determined by analyzing the relative abundances of certain chemical elements and their isotopes trapped within the rocks. The ratios of elements matched the measurements of the gases in the Martian atmosphere by the instruments on the *Viking* landers in 1976. The relative amounts of various chemical elements and isotopes in the atmosphere or rocks of a planet constitute a distinct mark of identity. Judged by its age as determined by radioactive dating, ALH84001 is unique among the Martian rocks found so far on Earth. While the other twelve meteorites range in age between 170 million and 1.3 billion years, ALH84001 probably crystallized from molten volcanic material on Mars 4.5 billion years ago, when the planets of the solar system were forming and the Martian crust was cooling. It is the oldest rock to be identified, not only from Mars but from any other planet, including the Earth. Unlike the Earth, where plate tectonics have buried all rocks older than 3.8 billion years old and it is almost impossible to find a rock of that age, Mars does not seem to have undergone similar geological processes and hence can provide evidence of a period missing from the terrestrial geological record (Mittlefehldt 1997).

How did ALH84001 reach Earth? On the basis of measurements of the effect of cosmic rays on various chemical elements within the rock combined with radiometric dating of its age, researchers have reached the conclusion that 16 million years ago this rock was ejected into space following the impact of an asteroid colliding with Mars. Orbiting the sun for most of this period, it finally struck the Earth and landed in Antarctica 13,000 years ago. Furthermore, about 4 billion years ago, when the newly formed solar system was at the end of the heavy bombardment period, the area of the Martian crust where the rock was located was struck by an asteroid from space, the impact of which melted part of the rock's content and created fractures and cracks within its matrix (Mittlefehldt 1997). In addition to these dramatic historical details, the extraordinary news delivered by the research team and published in the journal *Science* concerned the claim that this meteorite contains several lines of chemical and biological evidence implying the past existence of microbial life on Mars. The team was headed by the geologist and geochemist David McKay of the Johnson Space Center in Houston and included scientists of diverse specialties from the Johnson Center, Stanford University, McGill University, and the University of Georgia (McKay et al. 1996a).

The study of ALH84001 prior to 1996 had a history of its own, which will not be recounted here. I will however summarize the data presented by the research team in support of its contention of past microbial life on Mars and will discuss the scientific controversy as to their validity. During the few years since the first report on ALH84001 was published, a new astrobiology program jointly funded by NASA and the National Science Foundation has distributed sixteen small fragments of ALH84001 to investigators around the world in the hope of intensifying research on possible fossils in the Martian meteorite. According to a commentary in *Science* titled "Requiem for Life on Mars? Support for Microbes Fades," most scientists concluded from their investigation that as yet

ALH84001 did not reveal evidence of past biological activity on Mars. This general view was expressed at a NASA workshop organized to discuss the present situation and the prospects for research on Martian meteorites (Kerr 1998a:1398). Even McKay and his group have professed uncertainty about some of their original claims. Nevertheless, acknowledging that the experimental results reported by several laboratories are far from conclusive and presenting new data that may shed light on the similarity between the putative Martian microfossils and bacteria on Earth, McKay, the geochemist Everett Gibson, the microscopist Kathie Thomas-Keprta, and their colleagues claim that they are "more confident than ever" that ALH84001 does contain traces of ancient life on Mars (Kerr 1998a; Thomas-Keprta et al. 1998).

Despite the fact that the claim for remains of biogenic activity in ALH84001 is in great dispute, those facets of the ALH84001 episode that are relevant to our discussion will be examined here. The data produced by the various research groups are of interest in their own right. In addition, the investigation of the various features found in the rock and the heated debate over how to interpret them exemplify the nature of the scientific endeavor. First, it is obvious that despite the conflicting opinions, all researchers share the basic naturalistic outlook and its corollary regarding life on Mars as a valid possibility. Concerning the empirical issues, it is important to notice that unresolved questions are admitted by the researchers as such, and so are high hopes that have failed to materialize, at least for the time being. McKay acknowledges that he and his colleagues have not solved the question, which has proved to be much more complicated than previously anticipated. He does not foresee a clear resolution of the case of ALH84001 for the next five or ten years (Kerr 1998a:1398). At the same time, the persistence of McKay's team in its original contention despite the harsh criticism addressed against it clearly transcends the empirical issues involved and demonstrates the sociology of science at work. A great deal is at stake here in addition to the major question being addressed. As Richard Kerr noted in his News Focus column, with $2.3 million in funds from NASA and the National Science Foundation, "ALH84001 has become the most intensively studied 2 kilograms of rock in history" (Kerr 1998a). Moreover, the claims of possible ancient life on Mars have influenced NASA programs for the exploration of the solar system and beyond and the search for extraterrestrial life. Obviously, the controversy about various empirical details in the rock touches on highly crucial issues. Though money, ambition, and politics are all involved in this project, the investigation of the Martian meteorite also manifests one of the fundamental characteristics of scientific work, rightly described by the astronomer Donald Goldsmith as "an organized skepticism" (Goldsmith 1997a:226). This basic feature of the intellectual activity of science calls the bluff of the new creationist attempt, discussed in the last chapter, to portray the scientific study of the origin and evolution of life as driven by dogmatic faith.

The Biological Interpretation of the Meteorite Data

Getting back to the specifics of ALH84001, the reasoning leading the researchers to consider a biological interpretation of different features in the rock should first be reiterated. This reasoning draws on evidence of the presence of appropriate conditions for life on ancient Mars. In the 1970s, the *Mariner* and especially the *Viking* orbiters sent back to Earth images indicating that in all probability liquid water flowed on the surface of Mars in the past. Parts of the Martian surface that are several billion years old show features resembling river systems on Earth, including tributaries and lakes into which the rivers seemed to have flowed. Most of these valley networks date back to the end of the late bombardment, but some of them are much younger, pointing to the possibility of episodic floods during the history of Mars. In addition, there is evidence of volcanic activity throughout most of Mars's history manifested on the planet's surface by lava flows and flood basalt and the presence of several large volcanoes. It is assumed that liquid water was made possible in the past by a greenhouse effect caused by a much thicker atmosphere than now exists or, alternatively, by internal heat (Kasting 1997a:294–297). Even in the relatively warmer climate of the past, it is highly probable that water flowed on Mars under a thin layer of ice because of the planets's low average temperature. The internal heat sources point to the possible presence of subsurface hydrothermal systems that could have provided an environment similar to the hydrothermal vents on Earth, where organic material is postulated to have been produced and life might have originated (McKay 1997; Jakosky 1997:12–14).

Following various processes that took place between three and four billion years ago, the thicker older atmosphere, composed mainly of carbon dioxide and nitrogen, gradually disappeared. The probable causes were impacts by asteroids that blew most of the atmosphere off the planet, the dissolution of atmospheric carbon dioxide and nitrogen in water to form carbonates and nitrates, and, on a smaller scale, the escape of lighter atoms like hydrogen from the upper atmosphere. Though for several billion years Mars was still able to maintain liquid water under ice-covered lakes and later transiently in porous rocks, about a billion and half years ago the atmospheric pressure dropped even further to the point where no liquid water could have existed on the surface (Owen 1997; McKay 1997:270–271).

The specific lines of evidence supporting the hypothesis of past life on Mars presented by the ALH84001 team were the following: upon examining segments of the rock, the researchers found substantial amounts of carbonate minerals, previously detected in the other Martian meteorites only in trace amounts. These carbonate minerals form disc-like globules up to 250 microns in size (one micron equals one millionth of a meter) along the fractures in the rock and in pore space. Trying to establish the age of the carbonate globules proved to be a very difficult experimental problem, resulting in estimates ranging from 3.6 to 1.4 billion years, depending not only on the dating method but also on the specific hypothesis as to the mechanism of carbonate formation. It seems certain, though, that the globules are younger than the rock itself. On Earth, carbonates have usually formed underwater, and most of them are the result of biological processes like the accumulation of shells slowly producing limestone. Thomas-Keprta, an expert on scanning and transmission electron microscopy, special high-resolution techniques that reveal structures of a few nanometers (1 nanometer equals one-billionth of a meter), found the globules to be layered with fine grains of other minerals and to contain on their outer rims the magnetic minerals magnetite (iron oxide) and pyrrhotite (iron sulfide), and probably other iron sulfides (Goldsmith 1997a: 63–82).

In their *Science* paper, the researchers acknowledged that the occurrence of the fine-grained carbonate, iron sulfide, and magnetite phases could be explained by inorganic processes. In terrestrial geochemistry, however, each of these processes usually requires specific conditions that are incompatible with the others. Weighing the environmental circumstances under which the different minerals seem to have precipitated in the fractured Martian rock, the team concluded that simultaneous precipitation did not seem plausible in simple inorganic models, adding cautiously that "more complex models could be proposed" (McKay et al. 1996a:927). According to Hoyatollah Vali of McGill University, an expert on minerals produced on Earth by bacteria, the complex mineralogy found in ALH84001 might be the result of biogenic processes. Terrestrial bacteria produce magnetite and pyrrhotite grains similar to those found in the Martian rock, and co-precipitation of the two minerals within bacteria has been reported. Moreover, the researchers pointed out that magnetite particles in the rock "are similar (chemically, structurally, and morphologically) to terrestrial magnetite particles known as magnetofossils," which are fossil remains of bacteria found in a variety of sediments and soils (McKay et al. 1996a: 928).

Under high-resolution electron microscopy, examination of the surfaces of the carbonate globules revealed aggregates of tiny ovoid structures the longest being 100 nanometers (about four millionths of an inch) (Goldsmith 1997a:70–75). The team considered the possibility that these structures may have resulted from inorganic processes, like the partial dissolution of the carbonates. However, there is no example in the terrestrial geological record of such a phenomenon. The possibility of artifacts or contamination (for instance, dust grains) from handling the samples in the laboratory was ruled out by various control experiments, such as exposing a lunar rock chip to the same procedure. Finally, the possibility that these structures are the remains of past life on Mars was raised. The tube-like structures closely resemble terrestrial microfossils, though they are more than an order of magnitude smaller than the known fossilized remains of bacteria. Specifically, the researchers pointed out that the "ovoid features are similar in size and shape to nanno-bacteria in travertine and limestone. The elongate forms resemble some forms of fossilized filamentous bacteria in the terrestrial fossil record" (McKay et al. 1996a:928).

McKay and his team acknowledged that the origin of the carbonate globules was a key element in the interpretation of the Martian meteorite. While an inorganic process of formation under high temperatures was strongly supported by some researchers, McKay's group maintained that their analyses indicated a possible formation under low temperature, which was compatible with the hypothesis that the globules were the products of biological activity. This controversy over temperatures, as will be shortly described, is a major bone of contention among researchers. The hypothesis that the globules attest to the presence of ancient life in the rock, argued the team, could explain many of the observed features of ALH84001 discussed so far. It could also account for another finding reported by the researchers, the abundant presence of large organic molecules, polycyclic aromatic hydrocarbons (PAHs), on the fractured surface of the meteorite.

PAHs consist of two or more interconnected hexagonal rings, each made of carbon atoms attached to hydrogen atoms. These compounds are abundant on Earth, not in living organisms but as products of their degradation. In addition, PAHs form when organic fuels are burned and hence are produced by power plants and car engines. There are strong indications that these complex organic molecules are ubiquitous in space, in interstellar grains, and in interplanetary dust particles, representing up to 20 percent of the cosmically available carbon (Ehrenfreund 1999:1123–1124). Carbon compounds found in meteorites contain significant amounts of these macromolecules (Pendelton 1997:72–73; Snyder 1997:116). PAHs were identified in ALH84001 by Richard Zare of Stanford University, a leading organic chemist specializing in the detection of chemical compounds by laser chemistry and an expert in the study of interplanetary dust particles, by Simon Clemett also of Stanford, and other colleagues. The sophisticated laser techniques used at the Stanford "Zarelab" made it possible to separate and identify very small amounts of material without contamination by touching (Goldsmith 1997a:82–94). McKay, Zare, and their group claimed that the PAHs were indigenous to the rock, ruling out by various control experiments the possibility of contamination upon transportation to the laboratory and handling. They further argued that in all probability the PAHs are not the result of a contamination process during the long stay in Antarctica, since other meteorites found in the same area do not contain PAHs. Moreover, the highest concentrations of these organic molecules were detected inside the meteorite and not on its surface, and were found within the carbonate globules. The researchers also concluded that the PAHs were not formed in space and incorporated into the meteorite there. The PAH distribution in each of the extraterrestrial materials is different, reflecting the different environments in which these organic molecules were formed and subsequently evolved. The distribution of the various kinds of PAHs in ALH84001 is different from the profile identified in other extraterrestrial materials, including meteorites containing organic compounds, the carbonaceous chondrites (McKay et al. 1996a:924–926).

Though several scientists agreed with the claim that the PAHs found in the Martian meteorite are indigenous to the rock, the further claim made by McKay and his colleagues that these organic molecules are the products of degradation processes of ancient microorganisms is highly controversial. Most critics maintain that the presence of PAHs can be easily explained by various inorganic mechanisms (Anders 1996). The research team focused on the spatial correlation between PAHs and carbonates, suggesting that since carbonate is not known to be a special material for concentrating PAHs from an aqueous environment, it seems reasonable to assume that the PAHs and the carbonates formed at about the same time. Since a highly plausible inorganic process by which PAHs tend to form involves the exposure of organic substances to high temperatures, a key question is, At what temperatures did the carbonates form? Formation at low temperatures, which according to McKay and his colleagues is compatible with their data, could favor a biotic pathway, whereas high temperatures will tip the balance toward an abiotic scenario (Clemett and Zare 1996). At the same time, it has been pointed out by various scientists that whether the PAHs originated on Mars from the decay of bacteria or in an inorganic process from other organic molecules, the data from ALH84001 seem to provide evidence of the presence of organic matter on Mars (Jakosky 1997:15; Bell 1996). This is of particular interest in light of the negative results obtained in 1976 by one of the instruments aboard the *Viking* landers, which determined that there are no organics in the Martian soil (McKay 1997:264).

At the end of their 1996 *Science* paper, David McKay and his colleagues summed up the different lines of evidence that they found compatible with the past existence of life on Mars: a rock penetrated by fluid along fractures, in which carbonate minerals were then deposited; images produced by high-resolution electron microscopy that reveal carbonate globules and features resembling terrestrial microorganisms or microfossils; magnetite and iron-sulfide grains that resemble the products of terrestrial microbial systems; and the presence of PAHs associated with the carbonate globules. None of these observations, they said, "is in itself conclusive for the existence of past life. Although there are alternative explanations for each of these phenomena taken individually, when they are considered collectively, particularly in view of their spatial association, we conclude that they are evidence for primitive life on early Mars" (McKay et al. 1996a:929).

Critical Voices

Even before the *Science* paper was published, during the first press conference, dissenting voices were heard. NASA and the McKay team invited William Schopf, a paleontologist at the University of California, Los Angeles, and a leading expert on ancient terrestrial microfossils, to present critical comments on the findings. Schopf maintained that though the Martian origin and the age of the rock and also the presence of carbonate globules and PAHs were well established, the evidence of remnants of biological activity was still preliminary. The different features within ALH84001 could have been of nonbiological origin. Schopf quoted the late Cornell University astronomer Carl Sagan, saying that extraordinary claims require extraordinary evidence. In Schopf's view, the evidence brought by the meteorite team to justify their extraordinary claim of past life on Mars was far from being extraordinary and convincing. The 3.5-billion-year-old microfossils found by Schopf in ancient sedimentary rocks on Earth manifest cell-like structures enclosed in a cell wall and are grouped in what looks like a chain of cells. Some of the cells appear to be in the process of division (Schopf 1983). Finding a group of cells that show a cell wall, claimed Schopf, would provide much stronger evidence than that provided so far. This line of criticism, which assumes that life on Mars, if it ever emerged, was morphologically similar to life as we know it on Earth, should be kept in mind for further discussion.

Since the publication by McKay and his team of their results, many other groups have had a chance to examine pieces of ALH84001 and present their own conclusions. Controversy centers on several basic issues concerning the interpretation of the data, and, as already indicated, most researchers do not find the evidence of past biological activity convincing. A major issue concerns the size of the putative fossils in the meteorite. The ovoid structures discovered by the McKay group were 0.1 micrometer or less in diameter. The smallest known terrestrial bacteria are 0.2 micrometer in diameter. Furthermore, theoretical considerations point to a minimal size limit based on the number and size of macromolecules necessary for a living cell. It is agreed upon that each cell needs a membrane to separate it from the external environment and a basic molecular apparatus to run its metabolism, information storage, and replication. The conclusion drawn by most researchers is that the putative fossils in ALH84001 are too small to have been alive (Maniloff 1997; Nealson 1997; Psenner and Loferer 1997). They also reject the reference to terrestrial "nanobacteria" comparable in size to the Martian "nanofossils," reported to have been found in different mineral deposits (Folk 1997). It is not clear yet whether these earthly entities are indeed alive. There is no doubt that the analysis of details in ALH84001 at the micron and nanometer level poses difficult experimental and theoretical problems. McKay and his colleagues argue that as microscope technologies improve, smaller and smaller bacteria may be discovered on the Earth, and analysis of nanofossils in the meteorite from Mars, revealing more inner structures, might become possible. They also raise the basic question as to whether life on Mars might have developed differently from life on Earth: "Conditions on Mars may have favored the evolution of very small microorganisms as compared with typical terrestrial ones" (McKay et al. 1996b).

A further blow to the claim of nanofossils in the Martian rock came from another study presenting nanometer-scale images of ALH84001. Using the same techniques as McKay's team, John Bradley of MVA Inc. and his colleagues contend that most of the putative microfossils are nothing more than microscopy artifacts, narrow ledges of minerals that look like fossil bacteria (Bradley et al. 1997). Thomas-Keprta, who conducted the original scanning electron microscopy study, argues in response that there is a need to distinguish between mineral structures and Martian microfossils; the latter are located on the rims of the carbonate globules (McKay et al. 1997). Even though Bradley concedes that some of the elongated forms in the rock could be Martian nanofossils, the majority, he says, are mineral edges or even magnetite grains (Kerr 1997b). The general feeling is that microscopy technology alone will not be able to resolve the controversy, and other evidence will be needed to settle the question of past life on Mars. At a recent scientific meeting, Thomas-Keprta announced that she had developed a technique that will allow her to dissect a single nanobacterium and examine its contents (Kerr 1997a).

At a recent workshop titled "Size Limits of Very Small Microorganisms," organized by the National Academy of Sciences, most participants agreed that unless a radically different biology is possible the size limit of about 200 nanometers is necessary for a cell capable of metabolism and replication. This limit is based on the minimum number of genes, 250, found to be shared by the most primitive organisms and on the requirement of ribosomes and other essential cellular features (Vogel 1998b). A dissenting voice belongs to Olavi Kajander of the University of Kuopio in Finland, who has isolated nanobacteria from blood and urine, most of which are between 200 and 500 nanometers, but some not larger than 50 nanometers in diameter (Kajander et al. 1996). Kajander raises an interesting possibility, arguing that the smallest nanobacteria might assemble to form a reproducing organism (Vogel 1998b). Moreover, some researchers contend that a simpler life form, which might have existed on Earth and Mars, containing ribozymes acting as both enzymes and replicating molecules could have been much smaller in size than life as we know it (Vogel 1998b). At the meeting, David McKay admitted that the objects found in ALH84001 and reported in the 1996 paper probably have to be ruled out as nanofossils because of their size. However, he still insists that the bacteria-like forms in the rock may be parts of Martian bacteria (Kerr 1998a:1399).

In their original paper, McKay and his colleagues noted that the use of different techniques leads to different conclusions as to the temperature at which the carbonate globules were formed. Citing measurements of the ratio of oxygen isotopes in the rock and several other criteria, the team suggested that the globules probably formed at low temperatures, under 80°C, and could thus be of biogenic origin (McKay et al. 1996a:928–929). Other research groups that studied the Martian rock came to the conclusion that the carbonate minerals were deposited at a high temperature, above 600°C. These scientists claimed that the structure of the magnetite grains found within the globules indicated that they could have formed only at a very high temperature, which would have precluded biological activity (Harvey and McSween 1996). A different opinion is that the carbonates could have formed through low-temperature processes that were nevertheless non-biological (McCoy 1997:558). In a paper published recently, a new theory postulating carbonate formation under low temperatures (room temperature or colder), during evaporation of surface water or groundwater on Mars lends support to the original claim made by the McKay's team (Warren 1998). Moreover, the paper suggests that previous theories predicting high-temperature formation of the carbonates are all seriously flawed. Most researchers believe now that the carbonates were deposited at a temperature under 300°C. The range runs from 0 to 300°, with no way to decide on a more exact temperature. Though the carbonate globules could thus have been produced by biological activity, there is no clear evidence that indeed they were (Kerr 1998a:1400).

The discovery of PAHs in the carbonate phase of ALH84001 by McKay's team and the characterization of these compounds as indigenous to the rock and distinct from PAHs from other extraterrestrial sources led to the claim that these organic molecules could have been produced by past Martian bacteria. Many researchers have countered that inorganic processes could have been responsible for the formation of PAHs on Mars and for their particular location and arrangement in the rock. Edward Anders of the University of Chicago, a specialist on meteorites, presented a harsh critique of the biological interpretation suggested by McKay and his team. The formation of carbonates, the arrangements of the other minerals within the globules, the magnetite and iron-sulfide grains, and the presence of PAHs, he argued, can be explained by inorganic processes with at least equal plausibility (Anders 1996).

A potentially more devastating critique claiming that some organic compounds in ALH84001 might be terrestrial contamination was published recently. Jeffrey Bada of the Scripps Institution and his colleagues have detected trace amounts of several amino acids in the carbonate component of ALH84001 (Bada et al. 1998). The researchers decided to check for amino acids in the rock because, unlike PAHs, they claim, these molecules play a crucial role in biochemistry. Furthermore, because of their chirality there is a way to distinguish between amino acids produced inside and outside a living cell. As noted previously, all amino acids in known organisms are of the L type, while any chemical abiotic process produces a mixture of L and D amino acids. Bada and his colleagues found that the composition of amino acids in ALH84001 is similar to that found in another Antarctic Martian meteorite, EETA79001, and to the composition detected in Allan Hills ice. A previous study indicated that the amino acids in EETA79001 originated from ice meltwater that percolated through the meteorite. Among the amino acids detected, the researchers found glycine, which is an achiral amino acid, a racemic mixture of DL-serine and L-alanine. Their conclusion is that in both meteorites the amino acids are of terrestrial origin. ALH84001, though, was found to contain low concentrations of endogenous D-alanine that might be from a biological source on Mars (Bada et al. 1998:364).

Another study, published by A. J. T. Jull of the University of Arizona and his colleagues, analyzed quantities of different carbon isotopes in the organic material and carbonate minerals of the two Martian meteorites. Using a gradual-heating technique, the researchers separated the organic fraction in the meteorites from the carbonate-rich fraction and measured the ratio of carbon-13 to carbon-12, as well as the carbon-14 content in the two fractions. The ratio between C-12 and C-13 indicates whether the samples originated by terrestrial organisms, which tend to use the lighter carbon isotope. The level of radioactivity of C-14 in the different fractions can also distinguish between a terrestrial and an extraterrestrial source, with a high-radioactive carbon indicating a terrestrial origin. The study concluded that the carbonate fraction in both ALH84001 and EETA79001 is extraterrestrial. The organic component in EETA79001, however, seems to be the result of terrestrial contamination during the Antarctic period. This is also the case for the great bulk of the organic material in ALH84001, which was characterized as a terrestrial contaminant (Jull et al. 1998). These results led the Bada group to maintain that since amino acids and PAHs are minor components of bulk organic carbon, both minor and major organic constituents in these two Martian meteorites are contaminants (Bada et al. 1998:365). It should be pointed out, however, that according to the Jull group there was a small component, less than 20 percent of the total organic material in ALH84001, that did not fit into the neat categories described above: in its C-14 content, it was not a terrestrial contaminant; its C-13 to C-12 ratio resembled organic carbon made by terrestrial organisms; and it was more heat resistant than organic material from Earth. Jull and his colleagues described this carbon component as "preterrestrial . . . of unknown origin" (Jull et al. 1998:368).

In response to these studies, McKay, Gibson, and Thomas-Keprta maintain that "earthly contaminants don't rule out Martian life" (McKay et al. 1998a). They claim that though the amino acids detected by Bada may be contamination from Antarctica, the fact that they are of the L type does not necessarily mean that they are from Earth. Referring to the basic question of the comparison between terrestrial and extraterrestrial life, they rightly note that "Martian organisms may have produced left-handed amino acids similar to those produced by terrestrial organisms." It is not surprising, they say, that during the long stay of ALH84001 in Antarctica processes of contamination took place. Nevertheless, following extensive testing, the Zarelab team at Stanford University concluded that PAH levels in Antarctic ice and in other Antarctic meteorites are much lower than those in ALH84001. Furthermore, PAHs are not as water soluble as amino acids and are less likely to contaminate a rock via meltwater. Admitting that the source of PAHs remains a mystery, McKay and his colleagues claim that their main interest focused on the PAHs associated with the carbonate globules, whose Martian origin was decisively established. The Bada group, on the other hand, did not determine whether the contaminant amino acids were on the surface of the rock or in the carbonate globules (McKay et al. 1998a).

McKay and his group feel that neither Bada's nor Jull's results invalidate their original hypothesis: "They only make the story more complex and challenging." Referring to the small component that defied easy identification reported by the Jull group and labeling it "the mystery component," they claim that this component "actually lends new support to our hypothesis." They raise questions as to whether this carbon-rich material includes PAHs and whether it was produced by the decay of Martian organisms. The study of Martian meteorites should continue in order to provide answers to such riddles (McKay et al. 1998a).

Among the several lines of evidence of past life in the Martian meteorite, the one that seems still to hold promise is the discovery of magnetite (Fe_3O_4) grains in the rims of the carbonate globules. At a NASA workshop held in November 1998 to discuss ALH84001, Thomas-Keprta presented evidence that about one-third of the magnetite in the carbonates was identical in size, shape, and chemical composition to magnetite produced on Earth by magnetotactic bacteria as internal magnetic compass (Thomas-Keprta et al. 1998; Gibson et al. 1998:2). Thomas-Keprta believes that the presence of this component "strongly suggests previous biogenic activity," since no inorganic source that will produce such grains is known. These grains of magnetite, she contends, are Martian biomarkers (Kerr 1998a:1400). Many researchers present at the workshop felt that at the moment there is no way to tell whether the magnetite grains are of biological origin or were produced by an inorganic process. Interestingly, at the annual meeting of the Geological Society of America held in October 1998, a report by the geochemists Brian Beard and Clark Johnson of the University of Wisconsin indicated that not iron oxide but isotopes of the element iron can serve as a biomarker in terrestrial rocks, in the magnetite grains of ALH8400, and in any other rock sample. The method is based on the preferential use by organisms of the lighter, rarer iron-54 over the more common iron-56. Unlike the isotopic fingerprints of carbon and oxygen, which are sensitive to temperature, the distribution of the heavier iron's isotopes is more robust. Beard and Johnson succeeded in overcoming a technical obstacle that had so far prevented the analysis of iron by inventing an ingenious new method of analysis (Kerr 1998b:1807–1809). Measuring the fingerprints of iron isotopes in the magnetite grains might shed light on the unresolved problem of ALH84001.

At the 1998 and the 1999 annual Lunar and Planetary Science Conference in Houston, David McKay and his team reported that they have found remains of fossilized microorganisms in another Martian meteorite called Nakhla. (This meteorite reached Earth in 1911 and was discovered near Nakhla, Egypt.) Using techniques similar to those used on ALH84001, the researchers claim that the evidence of past life in Nakhla is much stronger than in ALH84001. The round to ovoid structures detected in Nakhla and interpreted as mineralized bacteria demonstrate several characteristics similar to terrestrial bacteria. The size-range of 0.25–1 micrometers in diameter is well within the accepted size for most bacteria cells; some of the structures are attached to each other, reminiscent of dividing bacteria; and the particles' surface is textured and resembles mineralized biofilms commonly produced by bacteria. The team suggests that Nakhla may contain both an original Martian bacterial generation, embedded in the rock, and a generation that grew on the exposed surfaces after arrival on Earth. Preliminary work revealed similar features in another Martian meteorite that fell near Shergotty, India, in 1865. Emphasizing that the interpretation of data is still tentative, McKay pointed out that both Nakhla and Shergotty are much younger than ALH84001, the first being 1.3 billion years old and the second, 165 million years old. Remains of life in these rocks could reveal that life existed during much of the history of Mars. Anticipating the audience's scepticism as regards the new findings, McKay called not for support but for an open mind (McKay et al. 1998b; Holden 1999:1841).

All participants in the debate agree that the general question of past life on Mars has a better chance of being resolved by analysis of rock samples brought back from Mars. In a combined American-French project, a number of missions intended to gather samples from Mars and bring them back to Earth are scheduled for 2003, 2006, and perhaps also 2008 and 2010 (Morton 1999). There is certainly agreement that even though ALH84001 provides no direct proof of microfossils or past biological activity, it does support the hypothesis that conditions on ancient Mars were conducive to the emergence of life. The data that point to fluid penetrating the rock along fractures and pore spaces, which then became the site for mineral formation and the presence of organic material, are compatible with a wet and warm climate on Mars. As already mentioned, most researchers believe that the best explanation of the valley networks found on Mars is the extended flow of water in the past. This notion gained strength on the basis of the data collected on Mars and sent to Earth by the *Mars Pathfinder* and its wheeled microrover *Sojourner*, which landed on Mars on July 4, 1997. The images sent back to Earth, while open to other interpretations, suggest the possibility of an early environment rich in water in the Ares Vallis region, the landing site of the *Pathfinder*. The *Mars Global Surveyor*, launched in November 1996 and starting to orbit Mars in September 1997, was not yet in mapping orbit at the time of writing of this book, but is already sending data from atmospheric, topographic, and magnetic observations. Data from the *Surveyor* point to an ancient ocean on Mars (Kerr 1998b:1807). Among other objectives, *Surveyor* is supposed to locate areas where water might have flowed in the past or where water reservoirs might exist beneath the surface today. The *Mars Polar Lander*, launched in January 1999, and other robotic missions planned for the next few years are supposed to detect ice and water in frozen samples scooped from the ground and to analyze the samples' chemical content. They will gather information on meteorological and climate conditions, on subsurface compositions, and on geophysical and seismic networks of the planet. All these missions are intended to help direct future landers to locate rocks that might contain fossils of past Martian life (Vaughn 1998; Cunningham 1998; Albee 1998; Mars global Surveyor Web site).

The transport of alleged nanofossils on ALH84001 from Mars to Earth has inspired a renewed interest in the idea of panspermia. Unlike the panspermia theories of the turn of the century, which assumed the eternal existence of life in the universe, current versions claim that life might have emerged on one planet and then been transported to another one or indeed to many other planets. During the time of heavy bombardment, numerous asteroids struck the planets of the solar system, releasing rocks of different sizes into space. Hence, according to this conception, the discovery of remnants of biological activity in ALH84001 casts doubt on the idea of the independent emergence of life on Mars and Earth. Other possibilities are that life originally emerged on Mars and was then brought to Earth, that life originally emerged on Earth and was brought from Earth to Mars, and that life came to both Earth and Mars from someplace else (Von Hippel and Von Hippel 1996). Frank von Hippel of Columbia University and Ted von Hippel of the University of Wisconsin indicate that because the gravitational field of Mars is weaker than that of the Earth, a Martian origin is more likely. Furthermore, Earth is a more efficient sink of planet-to-planet material (Von Hippel and Von Hippel 1996). The speculation that life on Mars produced life on Earth is based on yet another consideration. About 4.5 billion years ago the Earth sustained a huge impact that supposedly formed the moon. Had life already emerged on Earth at this time, it would have been wiped out by this collision, which moreover could have stripped out the oceans and parts of the atmosphere. On Mars, which did not suffer such an enormous impact, life could have emerged and developed during this period (Kasting 1997a:285). Not necessarily ruling out these possibilities, most origin-of-life researchers nevertheless focus on the emergence of life on Earth, assuming that given suitable conditions, life could have also emerged on Mars.

"LIFE AS WE KNOW IT"

Behind the search for remnants of microbial life on Mars is the assumption shared by most researchers that liquid water was a common feature on Mars during part of its history. The speculation that there might be life on Mars today focuses on those undersurface niches where liquid water might still be found as a result of the melting of frozen water in the soil by geothermal or volcanic activity. Underlying this reasoning is the notion that life on Mars would be similar to life as we know it on Earth, for which water is a prerequisite. Terrestrial organisms are composed mainly of water and organic molecules, and the search for life on other planets is focused on the search for liquid water and organic molecules. The notion that extraterrestrial life might have the same basic chemical composition and biochemical processes as terrestrial life was put to the test in 1976, when two landers, *Viking 1* and *Viking 2*, were set down by NASA on Mars. The landers contained robotic biological laboratories designed to detect metabolic processes carried out by microorganisms in the Martian soil. The instruments and experimental setups had been tested and proved able to detect life on Earth under conditions as similar to the Mars environment as possible. Chosen for this purpose were the dry valleys of Antarctica, which, although the coldest and driest place on Earth, still harbor colonies of microorganisms below rock surfaces (McKay 1997:273; Goldsmith 1997a:181–190). The first results sent by the *Viking* landers seemed to indicate the presence of life and stirred great excitement: gases were exchanged between the soil samples and the local atmosphere in the presence of terrestrial organic material; organic molecules brought from Earth were oxidized by the soil samples, and radioactive carbon dioxide was incorporated into the soil (Sagan 1994:73).

The situation changed, however, when two of the three biological experiments were repeated under a variety of conditions and did not validate the presence of Martian microbes. The results of the third experiment were more difficult to interpret and bred controversy among the scientists involved in the project. Yet most scientists were not convinced that there was evidence of the existence of life, because an additional experiment designed to detect organic molecules in the Martian soil did not find any trace of them (McKay 1997:263–265). This disappointing result weakened hopes of finding current life on Mars. In light of the continuous fall of meteorites and interplanetary dust on Mars, which supposedly deposit organic material on its surface, the failure to detect such compounds was interpreted as the result of an active process of destruction of these molecules, perhaps by ultraviolet radiation. As for the positive data first registered by the *Viking* instruments, the generally accepted explanation is that they were the result of chemical inorganic processes rather than evidence of living organisms. It is assumed that oxidizing compounds, probably formed by ultraviolet radiation of the atmosphere, could have mimicked biological processes, producing the results registered by the *Viking* instruments (Sagan 1994:73–74; Heidmann 1997:79).

Not only the planning and interpretation of the *Viking* experiments, but also, as we have seen, the claims for signs of past life in ALH84001 were made on the basis of analogies with life on Earth: indications in the rock of the presence of water in the past, of organic molecules, and of minerals similar to those produced on Earth by bacteria. William Schopf's criticism of these claims focused on the need for a better-grounded morphological analogy with terrestrial microbes and microfossils. The lack of evidence of cell walls, of a range of sizes and shapes among the alleged nanofossils, and of any process of cell division was cited by Schopf as a major drawback. The recent suggestion, mentioned before, by David McKay and his colleagues that the Nakhla meteorite might contain mineralized bacteria is based on the claimed similarity between various features of the fossilized particles and their counterparts in terrestrial bacteria. It should be noted that if the presence of organic molecules of Martian origin in ALH84001 is verified, it could indicate that while the present oxidizing conditions on the surface of Mars seem to preclude organic molecules, atmospheric conditions in the past were very different and allowed for their stable presence.

Broadening the Scope—Other Possible Settings for Life as We Know It

In the search for life on Mars and other extraterrestrial bodies, scientists are guided by their acquaintance with life as known on Earth. Yet long-established ideas about the necessary environmental conditions for life and its survival have lately undergone radical changes. Microorganisms, sometimes called "extremophiles," have been discovered that thrive under extreme temperatures, acidity and alkalinity, salt concentrations, and pressure. Microbes have been found in boiling hot springs, in hydrothermal vents, and in pores of rocks in the hot pools of Yellowstone National Park. I have already mentioned the microbial ecosystem found in the porous subsurface of rocks in the dry valleys of Antarctica, where the mean annual temperature is −20°C. Another microbial ecosystem has been found there in perennially ice-covered lakes (McKay 1997:273). Geologists drilling below the Columbia Basin have found, at a depth of 1,500 to 3,000 meters, anaerobic bacteria living off volcanic rock and oxygen-free water, independently of solar energy. Some produce methane from hydrogen gas and carbon dioxide, and others produce hydrogen sulfide from hydrogen and sulfate. These microbes have been nicknamed SLIMES: subsurface lithoautotrophic ("rock-eating") microbial ecosystems (Digregorio et al. 1997:269). Extreme halophiles discovered in salt deposits in New Mexico are claimed to have lived inside salt crystals for 200 million years, maintaining a very slow metabolism similar to the reduced metabolic rate of active bacteria found deep in the frozen Siberian soil (Digregorio et al. 1997:272; Heidmann 1997:85–86). As already noted, the different extremophile species belong to both the Bacteria and Archaea branches of life, indicating that some of them might have evolved very early, close to the common ancestor at the root of the evolutionary tree (Woese 1987; Digregorio et al. 1997:269–272). These findings have inspired researchers to pursue the idea that the emergence of life might have taken place under high temperatures and pressures in the vicinity of hydrothermal vents.

Thus the definition of a suitable environment for life seems much less restrictive than in the past. New possibilities have opened up in the search for life, which might now extend to harsher domains on Earth and on other planets and satellites in the solar system. However, even within this extended conception, liquid water and the biogenic elements, in particular carbon, are still considered the absolute prerequisites for life (Gibor 1976; Postgate 1994: 251–252).

On Defining Life

Discussions of extraterrestrial life often include the complaint that looking for life as we know it outside the Earth involves a narrow-minded, chauvinistic attitude. We cannot base our definition of life, and hence of the conditions suitable for life, on one example, goes the argument. In order to decide whether life as we know it is indeed universal, it is then claimed, we should search for alternative life forms. Robert Shapiro and the late Gerald Feinberg, a professor of physics at Columbia University, advanced this point of view in several books and articles, concluding that "the discovery of one [alternative life form], with a different physical or chemical basis, would quickly settle the issue." (Shapiro and Feinberg 1990:251). Shapiro and Feinberg reject the limited search for planets where liquid water and carbon compounds are present as based on either the misguided "carbaquist" (carbon and water) assumption that water and carbon are uniquely fitted for life, or on the equally misguided "predestinist" assumption that our biochemistry is bound to arise throughout the universe following spontaneous chemical reactions. They suggest a definition of life that supposedly is independent of the local characteristics of life on Earth. Life is the activity of a biosphere, they say, and a biosphere "is a highly ordered system of matter and energy characterized by complex cycles that maintain or gradually increase the order of the system through an exchange of energy with its environment." Replication and subdivision into organisms and species are the specific strategies by which our own biosphere evolved. Other biospheres might adopt other mechanisms that maintain and increase their specific material and energetic order (Shapiro and Feinberg 1990:251–252).

The conditions for life to originate and develop in a particular location, according to Shapiro and Feinberg, are first a flow of free energy, in any form, shape or kind, and second a system of matter that through interaction with this energy can become ordered. This system can be either chemically or physically based. The third condition is a stretch of time long enough to build up the complexity "that we associate with life." On the basis of these conditions, they list several speculative suggestions for life forms very different from the "carbaquist" variant known to us. Among possible alternative bases for chemical life they mention life in liquid ammonia on a planet with temperatures near –50°C and silicate life on a planet where the temperatures are above 1000°C and silicates can serve as a liquid medium. Since they maintain that life need not depend on chemical reactions, they suggest several possible versions of physical life. Order might be maintained and increased, for instance, in plasma within stars, in solid hydrogen within an extremely cold planet, or through patterns of radiation within a dense interstellar cloud (Shapiro and Feinberg 1990:252–254).

This broad attitude toward the definition of life seems attractive compared to a more parochial point of view. Yet Shapiro and Feinberg and others who proclaim to be anti-chauvinistic while imagining different chemical or physical material bases for life in foreign locales still derive the formal characteristics of hypothetical extraterrestrial life from the principles of organization of life as we know it. This is definitely the case with the characterization of life as the activity of a biosphere and a complex system that maintains order through an exchange of energy with the environment. Viewing life on Earth as an ecological, global phenomenon, the realization that all organisms on Earth interact through processes in the atmosphere, the crust, and the seas is regarded by most biologists as a necessary complement of the view of the organism as an individual entity (Morowitz 1992:4–6). Many of the ideas derived from the Gaia hypothesis formulated by the British chemist James Lovelock, according to which life is the property of a whole inhabited planet, are commonly accepted. Lovelock reached this conclusion on the basis of his evaluation of life on Earth and its influence on the characteristics of the planet itself, for instance, on the chemical composition of the Earth's atmosphere. He claimed that the fact that the mixture of gases in the Earth's atmosphere is far from equilibrium is most convincingly explained by the activity of life. As a result, Lovelock suggested that the question of whether there is life on the surface of a planet, for instance Mars, can be decided by determining the composition of its atmosphere (Lovelock 1979). As we will see later, this idea has been adopted as a general research strategy by most bioastronomers. In the same vein, the thermodynamic definition of terrestrial life incorporated in Shapiro and Feinberg's general conception sees every organism on Earth as a system that maintains its self-organization by the constant supply of external energy.

The claim made by Shapiro and Feinberg, and echoed by many others, is that until we discover and characterize life from an extraterrestrial origin we have no way of knowing what is contingent and what is necessary for any living system (see, for example, Sagan 1994:71). Yet unless we speak of other forms of life that are, at least in some respect, variations on the earthly theme, this claim is circular. For how can we identify, not only practically but even theoretically, a specimen of extraterrestrial life as such if we do not have a clue what we are looking for? In other words, it is contended that without another example of life we cannot formulate a general definition of life, but without such a definition how can we find a completely alien life-form? It thus seems that relying in some way on "life as we know it" in dealing with alternative living systems is an inevitable procedure. Later discussion will deal with the question of whether this procedure is also justifiable.

It is a common frustrating experience of anyone who has tried to produce a definition of life that either the list of properties is too wide and applies also to nonliving systems, or else it is too narrow, excluding some living entities. A functional definition that focuses, for instance, on eating, metabolizing, and excreting, might apply also to an automobile but might not apply to a dormant seed. A thermodynamic definition that characterizes life as an open system, capable of self-organization through the flow of energy, will equally apply to several physical "dissipative structures," like a vortex of water or a burning flame (Sagan 1994:72; Emmeche 1994:34–36). The futility of any attempt to "define" life, to provide an exhaustive characterization of its nature, stems from the fact that "life," though laden with philosophical preconceptions, is obviously an empirical concept, and our ability to characterize it is conditioned by our specific experience and current scientific knowledge.

It is for these reasons, according to Immanuel Kant, that empirical concepts, unlike mathematical and philosophical concepts, cannot in fact be defined at all, but only made explicit. The purpose of making a concept explicit, Kant says, is to make use of certain characteristics "only as long as they are adequate for the purpose of making distinctions" (Kant 1965 [1781]:586 B756). Indeed, the different enumerations of the characteristics of life—focusing on its physiological, biochemical, genetic, thermodynamic, and evolutionary aspects—aim, first of all, to distinguish life from inanimate matter. It is important to realize that this is the case even when considered from an evolutionary perspective, which is the most comprehensive framework for the evaluation of life. For even though the emergence of life from inorganic matter is regarded as a continuous process, without a sharp dividing line between non-life and life, this very evolutionary process has brought about the emergence of unique biological features that did not exist before. Evolutionary explanations consist in pointing out both the connection to the old and the emergence of the new.

Every living system, as revealed particularly at the molecular level, is organized in a much more complex way than any ordered physical system known to us. The unique character of this complexity lies in the ability of an organism to maintain and reproduce its organization according to specific internal instructions, or information, manifested in specific macromolecules. This character is connected with the purposeful, functional nature of biological organization, in which each part serves the survival of the whole. The Shapiro and Feinberg definition is silent on those prominent biological features.

Despite debates over the relative importance of natural selection and other evolutionary mechanisms, most biologists agree that the unique nature of a living system, this informed organized complexity, could have resulted only from the working of natural selection during the long process of Darwinian evolution. This process, as I have shown, is based on the existence of systems that are able to reproduce, mutate, and reproduce their mutations. In previous discussions of the metabolic and genetic models of the emergence of life, I pointed out how the supporters of each of these models aim at explaining the emergence of systems capable of biological evolution. The mechanisms responsible for the emergence of such systems—natural selection, physicochemical constraints contributing to chemical selection, and the alleged principles of self-organization—are differently evaluated by the genetic and metabolic camps. There are differences as to whether natural selection could have worked in systems without a self-replicating genome. Yet all the models suggest various scaffolds that could have resulted in the evolution of a biological "arch." As such, they are all opposed philosophically to the notion of "intelligent design." Since accounting for the unique complexity and teleology of biological organization on the basis of chance is realistically impossible, the mutually exclusive creationist and evolutionist explanations are the only explanations possible. Hence it is the view of the majority of biologists that any living system, wherever it arises, is most appropriately characterized as the product of an evolutionary process. Richard Dawkins speaks of "universal Darwinism," according to which natural selection did not just happen to bring about adaptation and complexity of life on Earth but is a necessary feature of life anywhere (Dawkins 1992).

Justifying the "Carbaquist" Point of View

I have tried to show that our basic understanding of life on Earth inevitably guides our conception of life beyond the Earth, and that when looking for life elsewhere we are bound to look for signs similar to those familiar to us. I have conceded to Shapiro and Feinberg, and to other "anti-chauvinists," that this similarity could be in the realm of "form," not "matter": while living systems beyond Earth should possess a complex organization produced and maintained through the mechanisms of Darwinian evolution, they could be very differently composed. Indeed, the possibility cannot be ruled out that life-forms based on other chemical elements could have evolved in other physicochemical environments. Systems capable of reproduction, mutation, and the natural selection of adaptive mutations, on either hotter or colder planets than Earth, could be made of other genetic and catalytic materials in other media. In fact, there is no reason why all the possibilities enumerated by Shapiro and Feinberg should not exist somewhere in the universe. First, however, it can be confidently said that it is highly unrealistic to assume we will be able to locate them, even with the most advanced technologies now imaginable. Carl Sagan recommended, for practical reasons, a modest search strategy of looking for liquid water and organic molecules, though "such a protocol might miss forms of life about which we are totally ignorant." Sagan, however, believed that "if a silicon-based giraffe had walked by the *Viking*

Mars landers, its portrait would have been taken" (Sagan 1994:71). Indeed. But how about Shapiro and Feinberg's "reciprocal influence of patterns of magnetic force" (Shapiro and Feinberg 1990:254)? Would the portrait of this "life-form" also be taken?

More important, there are numerous chemical and physical reasons justifying a "carbaquist" position that envisions life elsewhere in terms of water and carbon compounds (Owen 1980). It is not because of chauvinism that the majority of researchers believe carbon compounds in a liquid-water medium provide the best material, compared to other chemical combinations, for life on other planets. Furthermore, new findings continue to accumulate concerning the abundance of organic chemistry in the universe, and the existence of extra-solar planetary systems, indicating that our carbon-water example might indeed be widespread in the universe. Having said that, it is still an open question how similar to earthly life other water-carbon systems might be. Will they use identical nucleic acids for information storing and reproduction and identical proteins for catalysis, or will they be based on different polymers? Will they rely on the other earthly biogenic elements nitrogen, sulfur, and phosphorus in addition to carbon, oxygen, and hydrogen? Will they use L amino acids and D sugars, or the other way around? These kinds of questions will be settled by future experience.

On the Properties of Water. Though our conception of the necessary conditions for terrestrial life has changed recently, there is no doubt as to the absolute requirement of liquid water for life as we understand it. Both organisms living in water and those that live on land but maintain an internal aquatic environment depend on water as a solvent in which organic compounds dissolve and chemical interactions take place. It is much more difficult to envisage biochemical processes taking place in a gaseous medium and close to impossible to see them happening in a solid medium (Sagan 1991:996). The outstanding properties of water as a solvent and many of its other unusual properties result from the fact that water molecules form weak chemical bonds, the "hydrogen bonds" among themselves and also with many other molecules, which consequently tend to pass easily into solution (Pauling 1959; Lehninger 1970; Fieser and Fieser 1961; Stryer 1995). Water molecules form hydrogen bonds because of the electric polarity of the water molecule, caused by the arrangement of electrons in the hydrogen and oxygen atoms. In addition to the effect of water as a polar solvent on other polar molecules dissolved in it, the interaction of water with non-polar molecules is also crucial for biological functioning. This is most notable in the complex effect of water on both polar and non-polar amino acids in an enzymatic protein chain, an effect that helps create the specific spatial structure of the protein, which is necessary for its activity as an enzyme.

The special arrangement of water molecules caused by the extensive hydrogen bonding among them gives rise to very unusual thermal properties. The temperature range within which water remains a liquid under Earth-like atmospheric pressure, between 0° and 100°C, is among the widest compared to other liquids. Water is also known for its high boiling and freezing points and for its extremely high specific heat, as a result of which bodies of water change their temperature very slowly in reaction to external changes and thus stabilize the environment. The remarkable fact that water reaches its maximum density at 4°C and then expands before freezing has enormous ecological significance on Earth and probably at extraterrestrial sites where life might be possible. Ice, being less dense than liquid water, floats on top of it and does not sink. As a result, when temperatures decrease, bodies of water become covered with ice but remain liquid beneath the surface, thereby enabling the survival of life. Were it otherwise, lakes and oceans would freeze from the bottom up. This remarkable property of water is responsible for the situation found, for instance, in lakes in Antarctica, where life survives under the cover of ice (McKay 1997:273–275). The same situation has been suggested for Mars, and it may also exist on Jupiter's moon Europa, which is covered with ice but is thought to harbor a subsurface liquid ocean.

Other solvents have been suggested as possible substitutes for water on other celestial bodies (Fox and Dose 1977; Gibor 1976). Hydrogen fluoride and hydrogen chloride might serve as good polar solvents, but since the cosmic abundances of fluorine and chlorine are very low, the possibility of a biochemistry based on them seems unlikely. Water, on the other hand, is in all likelihood the most abundant liquid in the universe. Since hydrogen, oxygen, carbon, and nitrogen are among the most abundant chemical elements in the universe, ammonia, made of hydrogen and nitrogen, should be considered a likely alternative solvent. Liquid ammonia is also a polar solvent, due to the presence of hydrogen bonds among its molecules, but not to the same extent as water. Furthermore, ammonia is a liquid between −78° and −33°C and is in a gaseous state on Earth or any Earth-like planet. Only on very cold planets could ammonia be considered a possible candidate, and on such planets chemical reactions would occur much more slowly because of the lower temperatures. Comparing ammonia's properties with those of water—for instance, specific heat and surface tension—ammonia is found to be less compatible with life than water is. In its solid state, ammonia, unlike water, does not float on top of its liquid but sinks to the bottom. Ammonia as a liquid medium would not allow life to survive under its ice (Sullivan 1970:118–119).

The first research to carry out an extensive comparative analysis of the physical and chemical properties of water (and of the carbon compounds, especially carbon dioxide) from a biological point of view was the Harvard biochemist Lawrence J. Henderson. Without knowing about the hydrogen bond, which was discovered in the early 1920s, he reached the conclusion that water and the carbon compounds compared to other alternatives are most suitable for life as we know it (Henderson 1970 [1913]). In light of his analysis, Henderson spoke of the "fitness of the environment for life," formulating his conclusions in terms that could indeed be interpreted as "predestinarian." Present-day evaluations, more attuned to the picture of evolution as a tinkering process, are more cautious in tone. Because of this evolutionary conception and more extensive scientific knowledge, the possibility of solvents other than water is considered more seriously. Furthermore, despite the many outstanding properties that qualify water to serve as a most suitable medium for life on Earth and elsewhere, it is nevertheless acknowledged, especially in the origin-of-life context, that the presence of water seems to raise problems for an emerging living system. Christopher Chyba, while speaking of water as "the stuff of life," reminds us that water greatly interferes with the linking of amino acids and nucleotides into chains, a crucial step in the origin of life. "Water-based life," he says, "must therefore fight a constant battle against destruction" (Chyba 1998:17). André Brack of the CNRS in Orléans, on the other hand, points out that in situations involved in the origin of life, because of its powerful hydrolytic properties, water could have blocked the main chemical pathway, thus opening other pathways that would have few chances to occur in an organic solvent. Brack also emphasizes that because organic molecules have a dual structure containing both hydrophobic and hydrophilic groups, water is the agent that drives the conformation of these molecules (Brack 1993).

On the Properties of Carbon. Like the argument for water, the case for carbon as the most suitable chemical element for life is based on both its cosmic abundance and its remarkable properties (Pauling 1959; Lehninger 1970; Fox and Dose 1977; Gibor 1976). Carbon is the major chemical element in macromolecules carrying information, nucleic acids, and in organic catalysts, proteins. The arrangement of electrons in the atom's outer shell is responsible for carbon's ability to bond with up to four other atoms. This enables it to constitute the backbone of long polymeric molecules and also to bind to side groups that are necessary for the polymers' functions. The fact that carbon does not form more than four bonds, unlike silicon and sulfur, which occasionally do, prevents crowding around the backbone, which tends to inhibit chain formation.

Silicon, the element most similar to carbon in the periodic table, which due to its outer shell's electronic configuration can also form four bonds, is often suggested as a possible substitute for carbon in an alternative biochemistry on another planet. However, not only is carbon ten times more abundant in the universe than silicon, but there is a marked difference between the strength of the chemical bonds involving carbon and silicon atoms. Silicon-silicon bonds are much weaker than carbon-carbon bonds and tend to break more easily. A stable polymer backbone made of silicon atoms cannot exceed thirty to forty atoms at room temperature, while carbon chains, in DNA for instance, contain millions of atoms. Only below −200°C can silicon form long-chain molecules. At the same time, carbon compounds containing carbon-carbon bonds can be easily induced to interact further, because the energy involved in the formation of carbon-carbon bonds is not very different from that for the formation of carbon-hydrogen, carbon-oxygen, or other bonds involving carbon. Silicon-hydrogen and especially silicon-oxygen bonds, on the other hand, are much stronger than silicon-silicon bonds. The fact that silicon interacts very readily with oxygen, which is a widely abundant element in the universe, also explains the great instability of silicon bonds, which are easily transformed into silicon-oxygen-containing polymers. These polymers, abundant in the Earth's crust in the form of silicates, are extremely stable and do not seem to be able to participate in a biochemistry similar to ours.

Neither the enormous number of carbon compounds nor the unique variety of carbon's combinations with hydrogen and oxygen, and also with nitrogen, phosphorus and sulfur, can be matched by any other element in the periodic table, and thus carbon is established as the favorite candidate for participation in the metabolism of every living system and the constitution of its major biomolecules. The enormous number and diversity of carbon-containing molecules is made even larger by carbon's ability to form not only linear chains but also rings and to participate not only in single chemical bonds but also in double bonds. This fact enables carbon compounds to participate in the process of photosynthesis.

Carbon's properties, compared to silicon, make it not only an optimal constituent of living systems but also a major factor in their external environment. Carbon dioxide, the major carbon compound in the Earth's atmosphere, is a gas at the temperatures of the Earth's surface, and its great mobility in the environment is further enhanced by its solubility in water. Due to carbon dioxide's absorption coefficient in water, which is nearly 1.0 at the ordinary temperatures of the seas, the amount of free carbon dioxide in the water is almost exactly equal to its amount in the air. This carbon compound always accompanies water, which accounts for its great external mobility and its participation in ecological cycles connecting the inorganic environment and the biosphere. The solubility of carbon dioxide in water also enables many of the most universal physiological processes. In addition, carbonic acid, the acid compound of carbon dioxide, and its salts, the bicarbonates, are extremely effective in maintaining the neutrality of the internal environment of an organism. It should be pointed out that since carbon-dioxide molecules are much less interactive with other molecules than are oxygen molecules, carbon dioxide should be present in any planet's atmosphere (Goldsmith 1997b:211).

In contrast to carbon dioxide, silicon dioxide is a solid on planetary surfaces and is insoluble in water. Shapiro and Feinberg mention silicate as an alternative base for chemical life. At a temperature above 1,000°C, silicates become liquid and "could serve as a basis for evolving chemical order." They cite as possible locales for the evolution of silicate life a planet close to its sun or even the molten interior of our planet (Shapiro and Feinberg 1990:253).

There is no reason to rule out such a possibility. Yet judged by the fact that carbon is about ten times more abundant than silicon in the universe, and because of its remarkable properties, it seems that the chances of carbon-based life in the universe are much higher than those of its silicon-based counterpart. To reiterate, the strongest case for carbon as the major biogenic element and also for water as both constituent of life and its environment, not only on Earth but in the universe, is based on the conception of organisms as complex, self-organized, self-maintaining, and self-reproducing systems that are the products of evolutionary processes. For any system to perform all these functions and to evolve—first and foremost, to be able to pass information from generation to generation—it must possess molecules that are large, complex, stable, and varied. Organic chemistry as known to us, based on carbon and water, provides a very sound basis for any such system.

At this point we should look again at the philosophical conception that underlies and motivates the argument presented by Shapiro and Feinberg. According to them, "The generation of life is an innate property of matter," a notion that they believe conflicts with the carbaquist point of view. They claim that carbaquist proponents are led by necessity to the conclusion that life is a very rare phenomenon in the universe, probably limited to the Earth. This conclusion, they find, is equivalent to the theological belief in the specialness of the Earth. In contrast, their own wide-scope perception speaks about the proliferation of different life forms in the universe, in harmony with different environments (Shapiro and Feinberg 1990:248–249). Following our discussion of water and carbon, we can ask whether Shapiro and Feinberg's argument serves their philosophical position.

I concur with most origin-of-life researchers that life indeed is not a rare phenomenon in the universe. Instead of saying that it is an innate property of matter, which I find to be a rather predestinarian expression, I would say that life emerged, on Earth and elsewhere, by physical and chemical processes resulting from the properties of certain chemical compounds under specific environmental conditions. From what we know of the properties of the different chemical elements, and under the assumption, strongly substantiated by the work of science, that physics and chemistry as we know them are universal, our examination indicates that the combination of the properties of the biogenic elements, specifically of water and carbon, is peculiarly suited to constitute systems we call living. It is highly probable that living systems are what they are not only because of these properties but also because these chemical atoms and compounds are cosmically available. Hydrogen, oxygen, carbon, and nitrogen are among the most abundant elements in the universe. Water is the most likely molecule to appear as a liquid on any imaginable planet. Carbon dioxide is also thought to be present on most planets that have atmospheres. Recent studies have found organic chemistry running rampant in outer space. Summing it all up, it seems likely that life based on carbon and water is anything but a rare phenomenon.

Organic Cosmochemistry. The realization that many organic molecules, some of which are biologically relevant, are not found exclusively on Earth is the result of discoveries of the last few decades. The significance of these findings can be better appreciated in light of theoretical and empirical developments in astronomy (see Smolin 1997:116–128). There is a close connection between the presence of organic material in space and the structure of galaxies, including our own, and the dynamic processes within these giant celestial systems. One of the major discoveries of astronomers has been that the space separating stars in a galaxy is not empty but is filled with an incredibly dilute interstellar medium, which contains clouds of gas and dust. The interstellar medium is closely involved in the cycles of birth and death of stars in the galaxy. Massive stars, several times the mass of our sun, explode as supernovas when they "die" and in this process disperse their material back to the interstellar medium. This material, which consists of heavy elements such as carbon, oxygen, iron, and silicon, contributes to the formation of dust grains. The major source of matter in the interstellar medium, however, is streams of dust grains ejected by stars and carried into space by stellar winds. It has been discovered that before they leave their circumstellar envelopes and start to wander in interstellar space, these grains are covered by water, methane, and ammonia ices (Heidmann 1997:24–25). From studies by Mayo Greenberg of the University of Leiden and others, it is well established that reactions on the surfaces of such grains can lead to the formation of complex organic molecules (Greenberg et al. 1993). Under the influence of ultraviolet photons from stars in the galaxy, active chemical radicals are produced on the grains' surfaces. When the grains approach a star, these radicals heat up and interact to form increasingly complex carbon molecules. During hundreds of millions of years, despite the incredible dilution of the interstellar medium and the intense cold in many of its regions, it is estimated that a typical molecular cloud, a few light-years in diameter, can produce quantities of complex organic molecules equivalent to the mass of our sun (Heidmann 1997:25)

In a brief review of current astrochemistry, Lewis Snyder of the

University of Illinois lists 110 molecular species already found in the interstellar and circumstellar clouds of gas and dust in our galaxy. This great variety of molecules, the majority of which contain carbon, was discovered by sophisticated spectroscopic techniques, including radio and infrared observations. The identity of the compounds was further established through comparison of the observed spectra with organic molecules produced in the laboratory. Many of the detected small molecules, such as hydrogen cyanide, water, ammonia, formaldehyde, and methane, are considered possible prebiotic participants in reactions producing the building blocks of life. Among the organic compounds found, such as hydrocarbons, amines, and alcohols, some contain long carbon chains made up of eleven, and perhaps even thirteen, atoms (Snyder 1997). There are also strong indications of an abundant presence in interstellar dust grains of PAHs, already mentioned in connection with ALH84001 (Puget and Léger 1989; Ehrenfreund 1999). It is crucial to realize that organic features similar to those found in our own galaxy have also been detected in other galaxies (Pendelton 1997: 73). The abundance of organic molecules in interstellar clouds is highly relevant to our discussion, because such clouds are the birthplaces of stars and planets. A star is formed when a cloud of gas and dust collapses under its own gravity, and the complicated process of its formation also involves the development of a disk of matter around the new star, a solar nebula, which eventually gives rise to a planetary system. "Thus," says the physicist and cosmologist Lee Smolin, "planets such as our own form from the same soup of gas, dust, and organic molecules that are the wombs and cradles of the stars" (Smolin 1997:126).

In a previous discussion concerning the nature of the primordial terrestrial atmosphere, I pointed out that there are strong indications that organic material reached Earth from outer space through the accretion of planetesimals and bombardment by comets and asteroids during the formation of the planet. Even those who doubt that interstellar organic molecules could have survived in the solar nebula and within comets and could have contributed to the origin-of-life process on Earth regard interstellar organic synthesis as highly relevant to the chemistry of the early Earth and other forming solar systems. According to this viewpoint, processes in interstellar clouds are seen as analogous to processes on the surface and in the atmosphere of a young planet and as indicative of the formation of organic molecules under various physical conditions (Snyder 1997:119). Most researchers, however, do consider comets and asteroids to be remnants of primitive material in our solar system in which interstellar organic molecules have survived. Yvonne Pendelton of the NASA Ames Research Center has found a strong match in the infrared patterns from interstellar clouds and organic material from meteorites (Pendelton 1997:69). A similar match has been observed between the spectra of cometary grains and interstellar clouds (Orò and Cosmovici 1997:100). As a result also of a comparison of the isotopic composition of interplanetary dust particles and certain meteorites and comets, these objects are believed to contain interstellar material (Tielens and Charnley 1997:24).

In recent laboratory experiments, PAHs were exposed to ultraviolet radiation under simulated interstellar ice conditions, producing various complex organic compounds. Some of these organic products are similar to compounds found in the carbonaceous fractions of meteorites that have reached the Earth (Bernstein et al. 1999). The researchers conclude that organic compounds that fell on the early Earth were more than simply a source of reduced carbon to be altered by terrestrial processes of chemical evolution. In all probability, these compounds had reached the Earth already in a complex form (Bernstein et al. 1999:1137).

Comets not only are the most primitive objects remaining from the solar nebula, but have also kept their primordial content in a frozen condition. It is well established that comets formed in the colder regions, at a very great distance from the sun, from icy planetisimals that contained water molecules and interstellar grains coated with organic material. Planetesimals not incorporated into the then-forming Uranus and Neptune were scattered and subsequently populated the Oort Cloud, the great cloud of comets that extends out to an enormous distance, tens of thousands of times the distance between the sun and the Earth (Owen and Bar-Nun 1995). When comets are disturbed in their orbits by a passing star, they plunge into the inner solar system. The organic content of comets has been investigated not only by infrared spectroscopy measurements made on Earth but also by probes aboard spacecrafts sent in 1986 to inspect comet Halley. The European Space Agency probe *Giotto* revealed that the comet nucleus may be made of as much as 25 percent organic matter, part of which may be in the form of organic polymers not yet clearly identified. A new kind of grain containing carbon, hydrogen, oxygen, and nitrogen, and

hence called CHON, was also discovered (Heidmann 1997:26–33).
A probe named *Stardust* was launched on February 1999 by NASA
on a seven-year mission to gather interplanetary and interstellar
particles and to encounter the comet Wild-2 in 2004. Several missions are planned to study in greater detail the nuclei of comets. A
fly-by Comet Nucleus Tour (*Contour*) mission is expected to be
launched in 2002 and to visit three different comets before it ends
in 2008 (Veverka 1998); another is the ambitious *Rosetta* mission
to comet Wirtanen, planned for 2011, with the aim of landing on
the comet nucleus to obtain samples of its surface, which will then
be returned to Earth (Heidmann 1997:38–42). Detailed information regarding the organic composition of the various comets to
be gathered in these projects will help determine the possible contribution of comets to the emergence of life on our own planet and
on planetary systems throughout the galaxy.

While the study of comets is very difficult, meteorites that
reach the Earth provide a handier source for the study of organic
material from space. As pointed out above, there are strong indications that, like those in comets, organic molecules detected inside meteorites originated from interstellar clouds that gave birth
to the solar system. There is even some evidence that the Murchison
meteorite that fell in Australia in 1969 contains silicon-carbide
grains that, judged by their isotopic ratios, were produced by a supernova event of a specific type. The shock wave produced by the
explosion might have triggered the condensation of the solar
nebula (Hoppe et al. 1996). The Murchison meteorite belongs to
the carbonaceous chondrites, which contain a variety of carbon
compounds. These meteorites are thought to be made of fragments
from the asteroid belt between Jupiter and Mars. More than five
hundred organic compounds have been identified in the Murchison
meteorite, including hydrocarbons, alcohols, carboxylic acids,
amino acids, sulfonic and phosphonic acids, amines, amides,
purines (adenine and guanine), and the pyrimidine uracil. Of the
seventy-five amino acids identified in the meteorite, eight are identical to amino acids found in proteins and are very similar in their
relative quantities to those obtained in the famous 1953 Miller experiment (No amino acid has been identified so far in interstellar
clouds. The search for amino acids in these regions of the galaxy,
which is focused now on glycine, is extremely difficult technically,
requiring special radio telescopes [Snyder 1997]).

Unlike the amino acids found in organisms on Earth, all of

which share the L-type chirality, researchers found the Murchison amino acids to be racemic, containing an equal mixture of L and D isomers. Their isotopic composition also points to their extraterrestrial, presolar origin (Cronin and Chang 1993). The suggestion that the origin of homochirality on Earth might be in space—that an extraterrestrial process produced chiral organic molecules in meteorites or comets, which then seeded the Earth—was discussed earlier. In conflict with previous studies, John Cronin of Arizona State University discovered L-amino-acid excesses in the Murchison meteorite and in another meteorite that fell in 1949 in Kentucky (Cronin and Pizzarello 1997). These findings are backed by a recent report that also claims to find an excess of L-alanine in the Murchison meteorite, lending support to the extraterrestrial-origin-of-organics idea (Bailey et al. 1998). According to Miller and Bada, who reject the idea that homochirality originated in space, even assuming that chiral amino acids did reach the Earth by meteorites, "once amino acids get to Earth, they would racemize in very short order" (Irion 1998:627; see Bada 1995).

Coming closer to home, evidence of the prevalence of organic chemistry on planets and satellites in the solar system is overwhelming. There is a consensus among origin-of-life researchers that exogenous material, organic molecules and water, delivered by comets and asteroids played an important role in the emergence of life on Earth. Based on observations as well as on theoretical considerations, the general belief is that such delivery occurred also on other planets (Sagan 1994). Of particular interest is the case of Venus and Mars, which were probably showered with organic material and water by icy planetesimals just like Earth and yet developed so differently. The differences in present conditions between the Earth and these planets resulted from the proximity of Venus to the sun—thus its much higher temperatures—and from later events responsible for the disappearance of most of the Martian atmosphere (Kasting 1997a:294–296). As for the outer planets of the solar system—Jupiter, Saturn, Uranus, and Neptune—and several of their satellites, they all manifest processes of organic chemistry dominated by the presence of methane and made possible by reducing conditions.

It is further claimed by astronomers that the delivery of water and organics by planetisimals could have occurred in planetary systems throughout the universe (Owen 1994). Our discussion of the cosmic prevalence of a variety of organic molecules lends support to the claim made by Cyril Ponnamperuma, a pioneer and leader of origin-of-life research, that "chemical evolution is truly cosmic in nature and [that] primordial organic chemistry may be described as organic cosmochemistry" (Ponnamperuma and Navarro-Gonzales 1995:122).

THE SEARCH FOR LIFE WITHIN THE SOLAR SYSTEM AND BEYOND

The Concept of the Habitable Zone

Astrobiology focuses on the search for habitable planets and satellites and, more ambitiously, for signs of life on such celestial bodies. As we have seen, for a number of reasons there is a general conviction that life elsewhere, being "life as we know it," requires for its emergence and maintenance liquid water, organic building blocks, and a solid surface, usually provided by a planet, on which liquid water can gather in substantial reservoirs (Mariotti et al. 1997:302–303; Goldsmith 1997b:38). Depending on a minimum atmospheric pressure, which in its turn depends on a specific combination of the planet's mass and temperature, liquid water on a planet's surface will last and not evaporate. Massive planets can retain an atmosphere at very high temperatures. Less massive planets, on the other hand, will retain an atmosphere only if their temperatures are sufficiently low but will lose it at higher temperatures (Goldsmith 1997b:43–44). A guiding concept in the search for habitable planets, one that takes into account all of these factors, is the "habitable zone" (HZ) of a given star, the distances from the star within which the prevailing temperatures on a planet allow for the presence of liquid water, or, more generally, of any solvent in the liquid state used by a life-form (Hart 1978). Originally, the question of whether a given planet falls within the habitable zone of its star was answered based on the luminosity of the star, its mass, its distance from the planet, and the planet's mass. Growing information about additional factors that influence the temperatures on a planet has contributed to radical changes in the concept of the habitable zone and has made it much more flexible (Chyba 1997a; Williams et al. 1997). In addition, stressing the element of a time window, long enough for life to emerge on a planet, the concept of a continuous habitable zone (CHZ), originally suggested by Michael Hart in 1978, was further elaborated by James Kasting (Kasting et al. 1993; Kasting 1997a:297). The question of whether liquid water was available in the past on the surface of Mars long enough for life to develop is a case in point. Another relevant question is whether the intervals between giant impacts on the early Earth that stripped out the oceans were of great enough duration to allow the emergence of life.

It is in reference to such questions that we can appreciate the significance of recent estimates that terrestrial life emerged in almost no time, geologically speaking. Having narrowed the time window considerably following the discovery of signs of life in rocks dating to 3.8 billion years ago, and realizing that the Earth was in turmoil until 4.0 to 3.8 billion years ago, some researchers even speak of no more than 10 million years "from primitive soup to cyanobacteria" (Lazcano and Miller 1994). Though it is impossible to draw general conclusions based on the single example of the Earth, the ease with which terrestrial life seems to have emerged is nevertheless suggestive. As put by Sagan, "It would be a truly remarkable circumstance if life arose quickly here while on many other, similar worlds, given comparable time, it did not" (Sagan 1994:74). A time span of 10 million to 100 million years opens many windows of opportunity for life to emerge on other bodies within the solar system on which liquid water might have lasted for restricted periods of time. The same logic applies to celestial bodies throughout the universe.

It is currently appreciated that in addition to the distance from its star, other factors may contribute to the presence of liquid water on a planet that would have otherwise remained outside its "conventional" habitable zone (Goldsmith 1997b:33–39). The most obvious additional factor to be taken into account is the greenhouse effect produced by certain gases in a planet's atmosphere. Water vapors, carbon dioxide, and to some extent methane are known to absorb infrared radiation emitted by a planet's surface, raising the temperature of the surface and the lower atmosphere. The very small amounts of carbon dioxide in the Earth's atmosphere raise the planet's temperature a few degrees. (Polluting the atmosphere with more greenhouse gases can cause a much more pronounced heating effect.) Even though Mars's atmosphere is one hundred times thinner than the Earth's, its predominant carbon dioxide does raise Mars's temperature a few degrees, although not enough to allow liquid water on its surface. One explanation offered for the possible existence of liquid water on the early Mars relies on the greenhouse warming caused by the hypothetical presence of small amounts of methane in the atmosphere. Methane on the early Mars, it is further speculated, might have been outgassed from hydrothermal vents or perhaps produced by microorganisms (Kasting 1997a:297). Venus, having the hottest planetary surface in the solar system (462°C), is an extreme example. This planet, whose thick atmosphere consists almost entirely of carbon dioxide, demonstrates a runaway greenhouse effect, a positive feedback process in which high temperatures caused by carbon dioxide evaporate water on the surface, heating the atmosphere even further through the presence of water vapors, and finally bringing about the total evaporation of all liquid water on the planet (Kasting 1997a:294). The presence of greenhouse gases in the atmosphere of a planet or a satellite, either in the solar system or beyond, may determine the temperature on this body and hence its habitability.

Another crucial factor determining a planet's temperature is geothermal heat, which results from radioactive decay within rocks. If the planet is large enough, this heat might contribute to processes of plate tectonics, leading to volcanic activity on land and beneath the seas. The discoveries on Earth of abundant populations of microorganisms in hot springs and deep-sea vents, and thus the possibility of terrestrial life flourishing independently of solar energy, provide evidence not only of an unconventional habitable zone but also of a densely inhabited one. Mars, not being as large as Earth, does not seem to have undergone plate-tectonic processes, but the huge volcanoes on its surface attest to previous heating activity, which might have resulted in the presence of liquid water in the past. It has been suggested that undersurface liquid water, and perhaps life, might occur on Mars today as a result of geothermal heat (McKay 1997:284–285). Similar processes associated with heat released by radioactive rocks might be taking place in other planetary systems in the galaxy and beyond.

Jupiter's Moon Europa—a Possible Abode of Life?

Internal heat on a planet or a satellite can be generated by the strong gravitational pull exerted by its neighbors. This phenomenon, tidal heating, is evident in the system of Jupiter and its moons, where it seems to create a habitable zone far from conventional limits (Goldsmith 1997b:34–35; McKinnon 1997:765). Sixteen moons have been identified around Jupiter, the largest planet in the solar system. The four largest moons, Io, Europa, Ganymede, and Callisto, which were discovered by Galileo Galilei in the seventeenth century, have been under the close scrutiny of the spacecraft *Galileo*, which reached Jupiter at the end of 1995 after a six-year voyage from the Earth. Being extremely massive, Jupiter exerts a very strong gravitational pull on Io, Europa, and Ganymede (the most massive moon in the solar system). Ganymede's outer neighbor, Callisto, seems to be undisturbed by Jupiter's pull. Io, the nearest to Jupiter among the big moons, is most strongly affected, besides being pulled to and fro by the gravitational fields of Europa and Ganymede. Io's tidal flexing creates such heat in its interior that sulfur jets are explosively released from its numerous active volcanoes. Consequently, Io is the most volcanically active body in the solar system. Europa, about the size of our moon, is covered by ice, but mounting evidence indicates that its interior is also geologically active as a result of tidal heating (McKinnon 1997:765–767; Kerr 1997c).

The idea that beneath the icy cover Europa might harbor a liquid ocean, combined with several other possible features, might make this moon a suitable place for life. The hypothesis of a subsurface ocean was first raised after the *Voyager* spacecrafts sent back pictures of the surface in 1979. Furthermore, this hypothesis seems to be substantiated by data recorded by the *Galileo* spacecraft, which flew just a few hundred miles from the moon. Several pictures show blocks of ice floating on apparently frozen slush and patches of ice that seem to have welled up from below. The fragments appear to have cracked, moved, and refrozen. In the view of several researchers, the *Galileo* images resemble the ice cover of the Arctic Ocean (Lowes 1998; Anderson 1998). Critical voices within the scientific community, however, have counseled caution. There are claims that the pictures from *Galilro* point to a deeper layer of softer ice. A deeper layer of water, some say, may have existed in the past but not now, and water might exist locally but not as a global ocean (McKinnon 1997:765). *Galileo*'s mission to explore Jupiter and its moons was supposed to end in December 1997 but was extended through the end of 1999 by approval of NASA and the U.S. Congress. Data collected by the time of writing of this book substantiate the claim for the existence of a water ocean beneath the ice. In addition, through the technique of near-infrared mapping spectrometry (NIMS), hydrated minerals have been discovered on the surface of Europa. These minerals were found by a solid-state imaging camera to be red and yellow, colors that are thought to indicate the presence of organic compounds (Anderson 1998:12–13).

The findings complement an earlier announcement that spectrometers aboard *Galileo* detected organic molecules on the surfaces of Ganymede and Callisto. Among the substances that might be responsible for the spectra observed, researchers mention a group of complex organic compounds, tholins (see below), that might have formed on the surfaces of the two moons by the interaction of simple carbon and nitrogen molecules under the influence of ultraviolet and particle radiation (McCord et al. 1997). Tholins have also been suggested as a major organic component of the atmosphere and surface of Saturn's moon Titan, another body in the outer regions of the gas-giant planets (Sagan 1994:75). Thomas McCord of the University of Hawaii and his group point out that the similarity of the spectroscopic data from Ganymede and Callisto suggests common origins or processes on the two moons. They also suspect, from comparisons with spectra of interstellar ices, that organic molecules might have reached the satellites around Jupiter from outside the system. These findings are also relevant for Europa and strengthen the possibility that a subsurface ocean, heated by gravitational pulls, might contain organic material and even life. Another planned mission to Europa equipped with sophisticated instruments could settle the question by measuring the thickness of the ice cover and the boundaries between ice and water with long-wavelength radar. There are also discussions of landing a robotic device on Europa to probe beneath the ice and test for life.

Saturn's Moon Titan—an Organic-Chemistry Laboratory

Reviewing sites in the solar system that may have a liquid solvent, the astronomer and science writer Donald Goldsmith rules out bodies with no surface, like Jupiter and the other gas-giant planets, and others with too thin an atmosphere, like Mars. He is left with two sites, Europa and Titan, Saturn's large moon (The possibility of subsurface liquid water and life on Mars was discussed earlier.). Even if Europa is discovered to have no liquid subsurface ocean, Titan still offers the possibility of a surface with a liquid solvent. At a temperature of –180°C, (–356°F), this hypothetical liquid is made of hydrocarbons. Liquids can exist on Titan despite its very low temperatures because of the moon's thick atmosphere, which produces sufficiently high pressures. Similar to Earth's, this atmosphere consists mainly of nitrogen, with a few percent of methane. Goldsmith thus speculates that an unfamiliar life-form, not based on water but on liquid hydrocarbons and thus still on carbon chemistry, might have originated and evolved on Titan (Goldsmith 1997b:44–46).

Titan's surface is veiled by an opaque, orange haze. The atmosphere is characterized by a complex organic chemistry based on the interaction between nitrogen and methane under the influence of ultraviolet radiation and charged particles. Simulation experiments under the presumed conditions on Titan, carried out by Carl Sagan and his colleagues at Cornell University and other research groups, produced gaseous organics and also solid particles that Sagan called "tholins" (from the Greek word meaning muddy). Optical measurements of the particles match the observational data from Titan's haze (Thompson et al. 1994). Sagan estimated that the processes by which tholins were produced in the upper atmosphere and then sank to the surface, in all probability continuing over the past four billion years, resulted in tens or hundreds of meters of tholins covering Titan's surface. From measurements of the moon's density and comparisons with neighboring bodies, it is believed by astronomers that parts of Titan's surface are covered with water ice. Several astronomers speculate that "islands" of such ice might rise above a liquid layer made of methane and another hydrocarbon, ethane (Heidmann 1997:46–47). According to calculations by Sagan and W. Reid Thompson, there is a 50-percent chance that liquid water resulting from the heat of impacts on Titan lasted for centuries at a typical surface location (Clarke and Ferris 1997:241).

Citing the narrowing of the time window for the origin of life on Earth, Sagan raised the question of whether a thousand years were enough for life to originate on Titan. He was motivated to ask this question by laboratory experiments in which tholins mixed with water produced amino acids, traces of nucleotide bases, and many other organic compounds (Sagan 1994:75). Thus, unlike Goldsmith, who considers the possibility of hydrocarbon life on Titan, Sagan entertained the idea of a more ordinary water-carbon life emerging on this extraordinary moon.

The very low temperatures on Titan and the fact that at such temperatures any chemical reaction is extremely slow do not seem to encourage Sagan's "thousand-years hypothesis." Yet many researchers who believe that because of its low temperatures Titan was not able to advance beyond the prebiotic stage nevertheless view Titan as "a model of the primitive Earth held in deep-freeze" (Heidmann 1997:44). Clarke and Ferris describe the many conditions required for prebiotic synthesis that exist on Titan and might have occurred on the primordial Earth (Clarke and Ferris 1997). While the prebiotic stage on Earth could not be preserved because of the geological transformations that the Earth underwent, organic processes lasting for billion of years on Titan may provide us with the missing information. Part of this information may be gathered in a close encounter with Titan's atmosphere and surface scheduled to take place in 2004. The *Cassini-Huygens* spacecraft—a composite of a Saturn orbiter, the *Cassini* spacecraft built by NASA, and the *Huygens* probe constructed by the European Space Agency—was launched in 1997 and is supposed to reach the Saturn system in 2004. The instruments aboard the probe are designed to analyze Titan's atmosphere and surface. The *Cassini* orbiter is scheduled to stay in the Saturn system for four years to report on a variety of its phenomena (Heidmann 1997:44–48).

The Search for Extrasolar Planets

Life as we know it emerged and evolved on a planet, on which large quantities of liquid water gathered and the surface provided an interface (on land or undersea), conditions considered to be conducive to prebiotic processes. Though various life-forms unbound by such restrictions have been suggested, it is commonly agreed that the search for extraterrestrial life amounts, first of all, to a search for extrasolar planets orbiting other stars (Butler and Marcy 1997:340; Goldsmith 1997b:43–44). Using a strong metaphor, the astronomer David Black has commented on this close connection between planets and life, saying that "planets can be thought of as cosmic petri dishes" (Black 1995). The discoveries of Copernicus and Galileo in the sixteenth and seventeenth centuries, and many other findings since then, displaced our sun as the center of the universe and established the non-special status of planet Earth. On the basis of these discoveries, following a mode of reasoning called the Copernican principle, astronomers believe that planets orbiting sun-like stars are a common phenomenon in the universe (Shklovskii and Sagan 1966:367; Beckwith and Sargent 1996:144).

Until 1995, this belief had not been substantiated by actual observation of an extrasolar planet. The situation changed dramatically in October 1995, when a planet orbiting the star 51 Pegasi was found by Michel Mayor and Didier Queloz at the University of Geneva (Mayor and Queloz 1995). During 1996 seven other extrasolar planets were discovered, mainly by Geoffrey Marcy at San Francisco State University and Paul Butler, currently at the Anglo-Australian Observatory in Epping, Australia, and the number keeps growing (Butler and Marcy 1997). As of April 1999, after the discovery of four more planets around solar-type stars by the Marcy team, the number of discovered planets was eighteen (see Marcy and Butler website). Astronomers are convinced that this is just the beginning of the search (see websites of missions to detect Earth-like planets). Most of the planets discovered so far have been observed near stars from 40 to 80 light-years away. Most of them orbit stars similar to the sun in mass, size, and temperature. Astronomers focus their search on planets at solarlike stars, wishing to find planetary systems similar to ours. Technically, planets orbiting colder or hotter stars, compared to the sun, are harder to detect. Extrasolar planets have not been detected directly, a technological feat not yet possible. They were discovered through Doppler-shift measurements based on the tiny gravitational effect planets exert on the velocity of their star, which is manifested in repeatable, periodic changes in the star's spectrum (Goldsmith 1997b:8–12).

The accepted model for the formation of planetary systems, based on observations of the solar system and stars in the galaxy, has led astronomers to entertain certain expectations regarding extrasolar planets. A planetary system is thought to develop out of a broad, flat disk of gas and dust, a protoplanetary disk, encircling a forming star. The protoplanetary disk is supposed to be the remainder of a cloud of interstellar gas and dust that collapsed, forming the star. The gradual accretion of dust grains in the disk produces planetesimals of different sizes, which then coagulate and merge, giving birth to planets of various sizes and masses (Glanz 1997). Many protoplanetary disks have been observed in our galaxy, the most notable being the disk around the star Beta Pictoris, first spotted in 1984. Recent observations with new infrared cameras and radio telescopes have detected dust disks around several stars, which are claimed to be a clear manifestation of the emergence of new planetary systems. This conclusion is based on measurements of holes in the middle of the disks, whose existence and size are best interpreted as resulting from the gravity exerted by already existing planets that cleared out the inner material. For example, according to the investigating team that detected a central cavity around the star Fomalhaut, based on estimations of the mass of the dust in the disk, the cavity resulted from the formation of rocky, Earth-like planets within the disk (Holland et al. 1998).

Though the presence of planets is only inferred, the evidence pointing to an abundance of protoplanetary disks around stars seems to
substantiate the accepted model for the formation of planetary systems (Beckwith and Sargent 1996).

The model is exemplified by our solar systrem, where the
gradual accretion of planetesimals produced the inner rocky planets and the much more massive outer planets, whose cores are made
of ice and rock. The causes of this differentiation are several: the
outer regions of the disk, spreading over more space, contained
more material that could be used in the formation of the outer planets; closer to the sun, heat prevented the formation of ice particles,
which farther away were a major contributor to the cores and
masses of the forming planets; once the massive cores of the outer
planets formed, they could attract and retain hydrogen and helium
gases, thereby creating the enormous masses of the gas-giant planets (Glanz 1997:1336). Until 1995, this envisioned formation of the
solar system provided the conceptual framework guiding planetary
science research. The discovery of the new planets seems to challenge this accepted model. Surprise came with the first planet detected around 51 Pegasi. It is a giant planet, of a mass half that of
Jupiter, but its distance from its star is one-twentieth of the Earth's
distance from the sun, or one-hundredth of Jupiter's from the sun.
At such a distance, the temperature on the planet is 1,000°C. Clearly
such a planet could not support life as we know it. The discovery
of several other "hot Jupiters," as these giant planets whose orbits
are extremely close to their stars have come to be called, followed
suit. Among the new planets only two seem somehow to fit previous expectations, being gigantic in mass but orbiting their stars at
greater distances than the hot Jupiters. Since one exhibits a very
bizarre elongated orbit, only the planet around 47 Ursae Majoris
looks somewhat familiar. The mass of this planet is more than twice
the mass of Jupiter, and its orbit is about twice the Earth's orbit
around the sun (Glanz 1997:1337; Butler and Marcy 1997). These
two planets might lie in a habitable zone as far as their temperatures are concerned, but being giant planets made of gas, they probably lack the solid surface required for life to develop.

Out of the eighteen planets discovered as of April 1999, sixteen are orbiting around their stars in apparent isolation. In April
1999 came the discovery of three extra-solar planets that orbit the
star Upsilon Andromeda—the first planetary system that may be
compared to our own. The closest planet in the Upsilon Androm-

eda system was discovered by Marcy and Butler in 1996. The other two were independently found by the Marcy and Butler team and by a team of astronomers from the Harvard-Smithsonian Center for Astrophysics in Cambridge, Massachusetts, and the High Altitude Observatory in Boulder. The first planetary system was detected in a survey of about one hundred stars out of two hundred billion stars in the Milky Way Galaxy, suggesting, as was pointed out by the astronomer Debra Fischer, that planetary systems reminiscent of our own are abundant in the galaxy (Marcy and Butler website). Yet the differences between the planets around the sun and those orbiting around Upsilon Andromeda are vast. Like the other planets discovered so far, all three are giant in mass (three-quarters of the mass of Jupiter, twice the mass of Jupiter, and four times the mass of Jupiter) and close in orbit to their star compared to the giant planets of our own system (six-hundredths, four-fifths, and two-and-a-half times the Earth's distance from the sun), making astronomers wonder how such a system might have been created (Marcy and Butler Web site).

Citing physical principles, astronomers insist that planets of Jupiter-like mass could not have formed at the distance of a hot Jupiter from its star. Such planets had to be formed not closer than about three to five times the distance between the sun and the Earth for ice particles to survive. As indicated above, all three planets around Upsilon Andromeda are inside this theoretical "ice boundary." A mechanism that would explain the presence of such giant planets so close to their stars was suggested in the 1970s by Scott Tremaine of the University of Toronto and Peter Goldreich of the California Institute of Technology. Trying to explain the dynamics of Saturn's ring system, they suggested that planets could migrate inward after being formed far away from their stars. This process of orbital decay results from the interaction between density waves induced by the forming planet in the protoplanetary disk and the planet itself (Glanz 1997:1337). With the discovery of 51 Pegasi and the other hot Jupiters, it became clear that in addition to the inward migration of these planets from their original sites, the fact that they stopped just before being engulfed by their stars had also to be explained (Lin et al. 1996). Discussing this point, Mayor and Queloz and their colleagues speak of the need for "an antagonist torque acting in the very vicinity of the star." One of the proposed sources of such an antagonist pressure is the tidal interaction between the fast-rotating star and the approaching planet, which will be opposite to the torque due to the planet's interaction with the protoplanetary disk (Mayor et al. 1997:318–320). In addition, several possible mechanisms have been suggested that could explain the eccentric orbits of a few of the recently discovered planets. The interaction between two giant planets orbiting the same star may have thrown one planet inward and the other into an eccentric orbit. The effect of the interaction between two stars in a binary system has been suggested as an alternative mechanism (Rasio and Ford 1996). Reporting on the discovery of the Upsilon Andromeda planetary system, Butler commented that accreting from planetesimals in the disk surrounding the star, planets engage in a "gravitational tug of war" that determines the planets' orbits. Such gravitational interactions between Jupiter-mass planets "can play a powerful role in sculpting solar systems" (Marcy and Butler website).

The nature of the newly discovered planets raises many doubts and uncertainties. It is not clear, especially as far as the planets in eccentric orbits are concerned, whether these are indeed planets or rather "brown dwarfs," objects made of gas that had too little mass to become stars but were still formed from a collapse of dust clumps in the nebula rather than through accretion from planetisimals like the planets. Some researchers, using another detection technique, claim that the 51 Pegasi system in fact consists of the main star and a dim companion star, not a planet (Glanz 1997:1338). The discovery of the triple-planet system seems to alleviate such suspicions. It is hard to imagine that the three bodies around the star are anything else but planets. Several difficulties challenge the Doppler-shift technique, among them the question of the calculation of planetary mass, depending on whether an orbiting object is observed edge-on or face-on. Moreover, it is crucial to realize that this technique is biased toward the detection of hot Jupiters, which, being massive and moving very fast and close to their stars, exert a more easily detectable effect on the velocity of the stars. Their short orbital periods (for instance, 4.23 days for the planet at 51 Pegasi) allow many orbits to be observed over a brief span of time. In all likelihood, more Saturn- and Jupiter-like planets will be detected in the near future by the same method and a few other indirect methods. The direct observation of a planet is extremely difficult to achieve since even the most massive planets are relatively small objects that have no light source of their own and are very difficult to observe and resolve apart from their star. Even the Hubble Space Telescope, which orbits the Earth at a distance of 300 kilometers, undisturbed by the atmosphere, and is equipped with a very powerful mirror, cannot see planets around nearby stars (Goldsmith 1997b:49–62).

While my discussion has focused on the recent detection of massive planets, it is the question of habitable Earth-like planets that is our major interest here and a major motivation in the search for extrasolar planets. Based on the mechanisms suggested for the inward migration and brake of the hot Jupiters, some astronomers believe that no Earth-like planet could exist as part of such planetary systems. Most of the giant planets found so far would have destroyed inner and smaller planets or removed them from the habitable zone. Alternatively, all the material that might have accreted to form the inner planets could have been swept by the migrating giant planets (Goldsmith 1997b:158; Lissauer 1997). Peter Nisenson of the Harvard-Smithsonian Center for Astrophysics noted, however, that the observations of the Upsilon Andromeda system "cannot rule out Earth-sized planets as well in this planetary system" (Marcy and Butler Web site). At the moment the detection of planets resembling the Earth in mass and orbit is a moot question. As indicated above, such planets cannot be detected by the Doppler-shift or other indirect techniques. Most remarkably, in 1991 Alex Wolszczan, a radio astronomer at Pennsylvania State University, detected three planets of Earth-like masses around a pulsar, a radio-emitting spinning star, 1,400 light-years away. However, since a pulsar is a neutron star left after a supernova explosion, such planets, formed very likely after the supernova event, are considered an oddity by astronomers (Wolszczan 1994).

To resolve a system containing a star and an Earth-like planet so that the planet can be observed separately from its star, and in order to observe planets in greater detail, astronomers plan to use interferometers, systems that increase their light-collecting diameter by combining observations from separated collecting instruments into a single image. Interferometers using both visible light and infrared radiation are modeled on elaborate radio-wave interferometers that have been used successfully by astronomers (Goldsmith 1997b:62–65). In May 1998 the first of a series of huge telescopes in Chile that together will form the Very Large Telescope (VLT) started to operate. The VLT, scheduled to be completed at the end of the century, will combine the light-gathering power of its separate components into the largest telescope on Earth. Within the next decade or two, astronomers expect to be able to use several similar interferometers to discern extrasolar planets the size of Jupiter and Saturn and perhaps even smaller ones the size of Uranus and Neptune, with larger orbits. Success will depend not only on the combination of observations from the various components of such a system, but also on the technology of adaptive optics, by which telescopes are continuously adjusted to compensate for changes in the atmospheric refraction of starlight, which would otherwise prevent any precise observation of planets (Goldsmith 1997b:195–200; Mariotti et al. 1997).

The astronomer Robert Angel of the University of Arizona, an expert on adaptive optics, claims that this technique, combined with the advantages of interferometry, could yield images not only of Jupiter-size but also of Earth-like planets (Goldsmith 1997b:201–202). Most astronomers believe, however, that correction of the atmosphere's refraction by adaptive optics is not enough. It will require a space-borne interferometer fitted with adaptive optics and sent to the distance of Jupiter, beyond the zone where interplanetary dust particles still reflect sunlight and emit infrared radiation. There is no chance of discerning an Earth-like planet around its sunlike star in an orbit similar to the Earth's using visible light, because the planet would emit no visible light of its own, reflecting only a tiny fraction of the light intercepted from the star. However, a space-based infrared interferometer would be up to the task. Planets emit their own infrared radiation, discernible from the star's infrared radiation. Obviously, such an interferometer is a thing of the future, depending on the development of sophisticated Earth-based instruments and space-based technologies (Mariotti et al. 1997). These in turn depend on much-debated political decisions that will insure international cooperation—most importantly, between NASA and ESA (European Space Agency)—and enormous financial outlays.

Scheduled for launch in 2005, NASA's Space Interferometry Mission (SIM) is designed to determine the positions and distances of stars several hundred times more accurately than any previous program. This accuracy, achieved on the basis of the technology of optical interferometry, will allow SIM to probe nearby solar-type stars for Earth-sized planets. The technique used by SIM will still allow only an indirect observation of the planets. However, space-borne infrared optical interferometry, combined with light collectors on separated spacecrafts, and the building of large optical collecting areas—technologies now under development—will eventually lead to the construction of the Next Generation Space Telescope (NGST). Hopefully, space-borne telescopes powerful enough to directly observe Earth-like planets and determine whether these planets are habitable and inhabited will become a reality in the next decade or two (see NASA's SIM Web site; Goldsmith 1997b:197–208).

Habitable and Inhabited Planets

It is currently estimated by astronomers, based on the number of solar-type stars in our galaxy and statistical inferences drawn from the recent discoveries of planets, that "many millions of sunlike stars in the Milky Way have planets" (Goldsmith 1997b:156). The discovery in 1998 of a Jupiter-size planet orbiting a low-mass "red dwarf," which belongs to the most common type of stars in the galaxy, might mean that the number of planets is enormously higher even than that (Marcy et al. 1998).

I mentioned earlier the dust disks around several stars that seem to indicate the presence of planets. Protoplanetary disks are much easier to detect than planets or planetary systems, and their estimated numbers have also led astronomers to conjecture a very large number of planets in the galaxy (Beckwith and Sargent 1996). The Copernican principle, or rather Copernican reasoning, supports the assumption that among these myriads of planets there are many that are Earth-like. As we have seen, the actual discovery of such planets cannot be expected in the immediate future, but models of the growth of planetary systems and the distribution of planets in disks around stars developed by George W. Wetherill of the Carnegie Institution and by James Kasting predict that at least one habitable Earth-like planet if not two should be found in a typical planetary system (Sagan 1994:77; Lissauer 1997:295).

Wetherill has claimed that the formation of the giant planets in our solar system and the consequent diversion of most of the comets into the far-off Oort Cloud reduced the number of possible impacts that could have stripped the inner planets of their atmospheres and also diminished greatly the danger for life on Earth from massive bombardments. Hence, the existence of Jupiter-size planets in Jupiter-like orbits, the discovery of which is a somewhat more realistic technological goal than the discovery of Earth-like planets, might be a necessary condition for the presence of inhabited Earth-like planets orbiting the same stars (Wetherill 1994).

Astronomers inclined to optimism believe that a space-borne infrared interferometer capable of observing Earth-like planets and examining their atmospheres spectroscopically is feasible in the next few decades. The infrared spectrum—specifically, its middle (thermal) region—is most suitable for detecting Earth-like planets and also for determining whether a planet is actually habitable (Mariotti et al. 1997:302–305). Furthermore, it can indicate whether such a planet is, in all likelihood, inhabited. In 1980, Tobias Owen of the University of Hawaii theorized that large amounts of oxygen in a planet's atmosphere provide a good sign of extraterrestrial life. Oxygen is a highly reactive gas tending to combine easily with reduced molecules like iron compounds present on a planet's surface. Reduced compounds provided afresh from a planet's interior by volcanic activity will rapidly exhaust any atmospheric oxygen unless it is continuously replenished. Biological processes are the best candidates for doing just that. Life based on carbon and water, synthesizing its organic building blocks most likely out of carbon dioxide in photosynthesis or some other process, releases oxygen gas into the atmosphere. Since non-biological processes that might release free oxygen—notably the dissociation of water in the atmosphere by ultraviolet radiation followed by the escape of hydrogen to space—cannot compensate for the removal of oxygen by interaction with reduced gases, the net presence of oxygen in an atmosphere is a very probable sign of its biogenic origin (Owen 1980).

This conclusion must be qualified. First, a planet similar to Venus, closer to its star than Earth is to the sun and losing a lot of water because of a runaway greenhouse effect, would be able to create a transient atmosphere rich in oxygen. This would result from the high concentration of water vapors in the atmosphere, their rapid dissociation, and the loss of hydrogen to space. Oxygen as a criterion for the possible presence of life is thus confined to planets well within the habitable zone of their stars. An oxygen-rich atmosphere on an Earth-like planet can probably be interpreted as a sign of life (Kasting 1997a:302–305; Mariotti et al. 1997:303–304). As a second qualification, the possibility of an inhabited planet with no oxygen in its atmosphere should also be considered. Free oxygen first appeared in the Earth's atmosphere about two billion years ago, whereas life emerged long before that. The atmosphere during this early period might have been rich in methane produced by bacteria or might have contained sulfur originating from a photosynthesis process based on hydrogen sulfide in volcanic regions rich in this compound. Despite this caveat, the detection of oxygen on a planet is still a much more reliable criterion for life, since the presence of sulfur might be easily explained by non-biological processes (Kasting 1997a:305; Goldsmith 1997b:212).

Could the infrared interferometer referred to above detect oxygen in the atmosphere of extrasolar planets? Unfortunately, it could not. Atmospheric oxygen does not absorb radiation in the thermal region of infrared wavelengths, but ozone (O_3) does. Ozone is formed by the effect of ultraviolet radiation on oxygen, and its abundance increases logarithmically with oxygen's abundance. Thus ozone is a very sensitive recorder of the presence of oxygen (Léger et al. 1993). Other gases identifiable in this infrared region are carbon dioxide and water. Carbon dioxide is an excellent indicator of the presence of an atmosphere on a planet, but it is also observed in the atmospheres of Venus and Mars, which are uninhabitable (the latter at least on its surface). As for water, measurements of a planet's temperature and its distance from its star, which can be taken by a space-based interferometer, can be used to determine whether the presence of water vapors in the planet's atmosphere also indicates the presence of liquid water on its surface. Finding liquid water on a planet would mean that it is habitable, and the combined presence of water and ozone would point to a high probability of life (Kasting 1997a:300, 305).

A 1993 issue of the journal *Nature* carried on its cover the question, "Is There Life on Earth?" Carl Sagan and his colleagues reported on the results of a novel project aimed at detecting life on Earth using a spacecraft designed to fly by other planets (Sagan et al. 1993). The *Galileo* spacecraft, bound for Jupiter, had to be assisted in its voyage by the Earth's gravity, a fact that allowed its instruments to observe the Earth for a brief period. This "search for life on Earth" was based on several criteria, the most notable of which was suggested by James Lovelock in 1965. Lovelock claimed that a necessary condition for the presence of life is "a marked departure from thermodynamic equilibrium," which cannot be adequately explained by non-biological factors (Sagan et al. 1993:716). Specifically, Lovelock mentioned the abundance of a reduced gas such as methane in an oxygen-rich atmosphere. Using a near-infrared mapping spectrometer (NIMS) aboard the *Galileo*, both a very large abundance of oxygen and detectable amounts of methane were found in the Earth's atmosphere. As pointed out in the *Nature* paper, it is known that since methane is quickly oxidized to carbon dioxide and water, "there should not be a single methane molecule in the Earth's atmosphere" at thermodynamic equilibrium. No serious candidate for a non-biological mechanism that would pump methane into the atmosphere to produce a concentration 140 orders of magnitude higher than the thermodynamic-equilibrium value in an oxygen-rich atmosphere could be suggested (Sagan et al. 1993:718). On the basis of this and other observations made by the *Galileo*, Sagan and his colleagues concluded that there are strong indications of life on Earth, a fact that encourages reliance on the capability of fly-by spacecrafts to detect extraterrestrial life. Though they claimed that the evidence for life on Earth was inferred "without any *a priori* assumptions about its [the Earth's] chemistry" (Sagan et al. 1993:720), it is clear that carbon-and-water chemistry, though not necessarily identical to terrestrial life, was assumed.

As pointed out by James Kasting, this *Galileo* experiment does not seem practical as a guide for a project in outer space. Detecting a gas like methane on a remote Earth-like planet will require a high-resolution spectrum, like the one used by the NIMS, with a much larger interferometer than the ones now contemplated (Kasting 1997a:299–300). Assuming optimistically that the scientific, political, and financial difficulties involved in the development of the necessary technologies can be overcome, a more realistic prospect seems to be the construction of a mid-infrared interferometer that can detect ozone, carbon dioxide, and water in the atmosphere of Earth-like planets at about thirty light-years away.

Unlike the optimists within the astronomical community, others tend to see the prospect of detecting gases of biological origin on distant planets as belonging to the distant future. This view is shared by the astronomer Jill Tarter, a leading SETI (Search for Extraterrestrial Intelligence) scientist, at the SETI Institute in Mountain View, California, who claims that the search for radio waves from an extraterrestrial intelligence carried on today is a much more reasonable method of looking for life and, by inference, extrasolar planets (Tarter 1995:9). This claim brings us to the complex subject of SETI.

On the Question of Extraterrestrial Intelligence

The question, Are we alone in the universe? has been raised by humanity throughout history. For the last four decades, a scientific answer has been sought in the growing field of astrobiology. The development of radio astronomy in the 1950s, which opened up the possibility of detecting signals from distant stars in our galaxy, made it possible to look for extraterrestrial civilizations that might have developed the technological means for interstellar communication. A turning point in the domain of SETI took place in 1959. An article published in *Nature* by the astronomer Giuseppe Cocconi and the physicist Philip Morrison of Cornell University suggested conducting a SETI experiment by listening to a specific radio signal, the frequency produced by the nuclei of hydrogen atoms. Cocconi and Morrison believed that a scientifically advanced civilization on an extrasolar planet, assuming the universality of basic scientific knowledge, might have chosen this frequency of the most abundant element in the universe to communicate with an advanced terrestrial civilization (Cocconi and Morrison 1959). In 1960, the American astronomer Frank Drake conducted the first SETI search, using the hydrogen-frequency band and monitoring two sun-like stars among the closest to our solar system. Since then, various SETI projects—some surveying the entire sky and "listening in" at the hydrogen-frequency channel, others targeting specific stars and listening in at thousands of channels; some looking for a signal purposefully transmitted, others aiming to detect unintentional "leaked" broadcasts—have been conducted all over the world (Heidmann 1997:11–114, 121–133). The growing computational power of computers is making these projects possible. So far, notwithstanding several false alarms, no signal that could be interpreted as coming from an intelligent source has been reported (Sagan 1994:76).

Until now, SETI depended on small amounts of observing time at radio telescopes built for other purposes. In the next few years a huge telescope designed to search for alien technologies is going to be built by the privately funded SETI Institute of Mountain View, California, and the Radio Astronomy Laboratory at the University of California, Berkeley. By combining the signals from more than five hundred small dishes, the new telescope will allow a search of the sky one hundred times more efficient than is possible with telescopes available now (Goldsmith 1999). Another ambitious project is SETI@home—implemented as of spring 1999—in which a sophisticated screen-saver program will allow hundreds of thousands of Internet-connected personal computers around the world to participate in deciphering SETI data from the Arecibo Observatory radio telescope in Puerto Rico. Based on the innovative idea of SETI scientist Dan Werthimer of the University of California, computer scientists David Anderson and David Gedye, and University of Washington astronomy professor Woodruff T. Sullivan III, recipients downloading the unique program onto their home computers will be able to become active members in the search for extraterrestrial intelligence (see Planetary Society's Web site).

Questions and problems surrounding extraterrestrial intelligence—how it should be defined, whether it is rare or common, whether the search for it is at all feasible or worthwhile, the various complex technical aspects of the search, the financial and political factors involved, the social significance of the search and the implications of a possible contact—all these issues and many more have been extensively examined and discussed in numerous books (Davies 1995; Dick 1996; Drake and Sobel 1992; Heidmann 1997; Jakosky 1998; Shklovskii and Sagan 1966; Sagan 1973; Zuckerman and Hart 1995). Only a brief discussion of points relevant to the main themes of this book will be added here to this growing literature.

I pointed out earlier that two different questions are involved in estimating the probability of the emergence of life on Earth and beyond. The first deals with the frequency of biogenic conditions on celestial bodies, the other with the chances for the emergence of life should these conditions be fulfilled. On the basis of the discussion in this chapter, the first question can be paraphrased as dealing with the number of continuously habitable planets in our galaxy and in the universe at large (this number in turn depends on the number of sun-like stars that have planetary systems including Earth-like planets). A growing body of empirical data suggests that the probability of biogenic conditions on planets beyond the Earth is high. The abundance and variety of organic molecules on extrasolar planets and satellites and in interstellar clouds and the existence of planetary systems in the galaxy clearly point in this direction. So far, Earth-like planets have not been detected, and the motivation to search for them, despite the huge obstacles involved, is fueled by the Copernican reasoning that our position in the universe is not privileged, that out of the many millions of planets estimated to exist in the galaxy, Earth-like planets within the habitable zones of their stars are certainly to be found. Providing a more complete empirical answer to the question of the prevalence of biogenic conditions in the galaxy and the universe depends on the continuation of astrobiological research.

The second question, dealing with the probability of the emergence of life when appropriate conditions prevail, is of a different kind, and as indicated before involves the core philosophical issue of the nature of biological organization. While even new creationists are ready to grant that the prebiotic synthesis of organic molecules out of inorganic material on the primordial Earth was possible by natural means, they regard it as impossible to imagine a natural mechanism that would be responsible for the self-organization of these organic building blocks into a functional biological system. Matter, they believe, cannot organize itself into "an irreducibly complex system" (Behe 1996:39) but requires the helping hand of a supernatural designer. I noted earlier that because of the complex nature of even the most primitive living system and the intricate interdependence among its components, its original emergence was viewed as a "veritable enigma" (Monod 1974:135–136) or as "almost a miracle" (Crick 1981:88) by prominent biologists. Origin-of-life research, on the other hand, is based on the philosophical tenet, strengthened by empirical data, that processes of self-organization have brought about the emergence of life without the agency of intelligent design or the chance association of molecules. The various models formulated by researchers in the field postulate different physicochemical processes that could have set the stage for the operation of natural selection and the development of metabolism and genetic information. These models embody the extension of evolutionary reasoning to the emergence of life.

The actual discovery of life, present or extinct, in a Martian meteorite, under the icy surface of Europa, or on an extrasolar planet will dramatically change our ability to draw statistical conclusions about the chances of life in the universe. But even before that, based on the philosophical Copernican and Darwinian tenets, researchers of the origin of life and astrobiologists expect life not to be confined to planet Earth.

The question of the probability of civilizations in the universe capable of communicating by radio signals obviously involves additional factors beyond those already considered in connection with the probability of habitable planets and the emergence of life.

Relevant variables include the number of inhabited planets on which at least one intelligent species has evolved, the fraction of such planets where a sufficient level of technological advancement has developed, and the duration of time within which such a technological society would keep communicating before becoming extinct or losing the motivation to communicate. All these factors were included in a famous equation formulated by Frank Drake in 1959 to estimate the number of detectable technological civilizations in the universe. Despite its highly speculative nature, the Drake equation has served since the 1960s as a framework for all deliberations on extraterrestrial life, measuring both our knowledge and our ignorance (Chyba 1997b:157–158).

For many astronomers and physicists, and for some biologists, the Copernican and Darwinian logic is applicable not only to the origin of extraterrestrial life but also to the development of intelligence on other worlds. Carl Sagan, who was among the leaders in the search for extraterrestrial life and intelligence, estimated in 1966 that our galaxy might contain a hundred million technological communities (Shklovskii and Sagan 1966:451). In a later publication, taking into account the estimated rate of catastrophic impacts that might destroy life on planets, Sagan suggested that the number of civilizations in the galaxy that are long-lived and have not developed technologies needed for detection and deflection of asteroids or comets is small (Chyba 1997b:162–163; Sagan and Ostro 1994).

Even if we insist on much more modest estimates, the notion that we are not the only intelligent species in the galaxy and the wish to verify it seem to be, explicitly or implicitly, the main motivation for the exploration of space. In this sense, Jill Tarter's suggestion that we search for habitable and inhabited planets via the search for radio signals has more than technical force. Citing several factors, the astronomer J. Richard Gott III claims that the number of radio-transmitting civilizations in the galaxy is smaller than 120 and a search within the thousand near-by stars has no chance of detecting any of them. Nevertheless, Gott is a strong supporter of SETI, pointing out that even if the chances of success in our "neighborhood" are small, an all-sky radio search could be successful (Gott 1995:182). Gott's support of SETI is based on the Copernican revolution, which taught us, he says, "that it was a mistake to assume, without sufficient reason, that we occupy a privileged position in the Universe" (Gott 1993:315).

Many scientists, however, including prominent evolutionary biologists, question the basic assumption underlying the SETI enterprise. In their view, because of the nature of biological evolution, human intelligence is a unique phenomenon that, in all probability, has not evolved on any other planet in the universe. If this conclusion is valid, the Copernican reasoning that our planet is not unique is in conflict with Darwinian reasoning. Ernst Mayr separates the issues of the origin of life itself and the evolution of the human species. While regarding an extraterrestrial origin of life as probable, he sees the incredible improbability of the evolution of intelligent life, even on Earth, as an example of the opportunistic and unpredictable nature of biological evolution (Mayr 1995:152–156). Mayr points out that "there were probably more than a billion species of animals on Earth, belonging to many millions of separate phyletic lines, all living on this planet Earth which is hospitable to intelligence, and yet only a single one of them succeeded in producing intelligence" (Mayr 1998:284). All along the way, from the origin of life to human intelligence, chance events intervened to determine the result, and even though, Mayr says, rudiments of intelligence are found in other species, only one kind on Earth has been sufficient to found a civilization (Mayr 1998:283).

By portraying human intelligence as a highly improbable and unique product of evolution on Earth, Mayr and other opponents of the notion of extraterrestrial intelligence (ETI) attempt to refute the argument of "convergent evolution" used by proponents of ETI. During the history of life on Earth, the argument goes, species that do not share close ancestry nevertheless developed similar molecular, physiological, morphological, and behavioral traits. These similar traits are adaptations, selected for their survival value, that developed along different routes to solve similar problems (Bieri 1998). The independent evolution of the extremely complex, similar, but far from identical eyes of vertebrates and cephalopods is a case in point (Dawkins 1986:93–95). So is the development of flight in birds, bats, and insects. Philip Morrison, a SETI pioneer, mentions an interesting (and more SETI-relevant) example of convergent evolution, the social behavior of elephants and sperm whales. Both species have big bodies and brains, live for many years, and have human beings as their only serious predator. Both live in bands of a dozen or so that include a few related older females and their offspring. "Most strikingly, each species has developed an unusual means of remote communication among families. The elephants project and attend to infrasound over five to ten kilometers across African woodlands, and the sperm whales send and receive sonar-like click signals over hundreds of kilometers across the high seas" (Morrison 1997:573).

Since convergence is common among species on Earth, why not assume that it could occur between earthly species, including *Homo sapiens*, and species on other planets? Proponents and opponents of the application of convergence to the evolution of extraterrestrial intelligence disagree over the definition of intelligence and the uniqueness of intelligence on Earth. Proponents argue that intelligence has an obvious survival value and has evolved more than once on Earth in whales, dolphins, and the great apes, and could have evolved following different pathways on other planets. Comparing brain organization and behavior in primates (including humans) and cetaceans (dolphins, whales, and porpoises) as a possible model for the comparison of humans and intelligent extraterrestrials, Lori Marino of Emory University reached the conclusion that despite significant differences in brain organization, there is "a surprising degree of behavioral convergence" due to similar kinds of selective forces (Marino 1997). Sagan contended that while the evolution of human-like intelligence beyond Earth is indeed highly improbable, the evolution of "the functional equivalent of humans" is possible (Sagan 1973; 1995). Opponents of this view acknowledge that intelligence on another planet might appear to be entirely different from that of humans on Earth, and furthermore that a rudimentary kind of intelligence is widely distributed in the animal kingdom. However, Mayr says, "the point I am making is the incredible improbability of genuine intelligence emerging" (Mayr 1998:284). Though we cannot dwell here on the question of the definition of intelligence and consciousness and their interrelatedness, it is clear that the debate concerns the nature of biological evolution.

In his book *Wonderful Life*, Stephen Jay Gould presents the idea of contingency as the fundamental characteristic of history in general and the history of life on Earth in particular. Historical processes, he contends, are not ruled by chance, since every step has a cause. However, no series of steps can be repeated or predicted, because it involves thousands of improbable events. "Alter any early event, ever so slightly and without apparent importance at the time, and evolution cascades into a radically different channel" (Gould 1989:51). As a result of the contingent nature of evolution, the development of human intelligence was extremely improbable. We have to realize, Gould states, that the origin of *Homo sapiens* was "a tiny twig on an improbable branch of a contingent limb on a fortunate tree" (Gould 1989:291). At the same time, Gould does not regard the possibility of bacterial life on ancient Mars as surprising or unexpected. The emergence of life on Earth or any other place where the appropriate conditions prevailed, he said in his 1996 article in the *New York Times*, can be seen as "a virtually automatic consequence of carbon chemistry and the physics of self-organizing systems." Yet "humans remain as gloriously accidental as ever."

In *Wonderful Life* and in other publications, Gould makes it clear that he does not view evolution as a random, chaotic, senseless process. Neither does he challenge the notion that many crucial aspects of life's history follow directly from general laws of nature. Not only does he see the origin of life as "virtually inevitable" given the conditions on the early Earth and physical principles of self-organization, he also claims that "invariant laws of nature impact the general forms and functions of organisms; they set the channels in which organic design must evolve" (Gould 1989:289). Yet Gould's contingency thesis and his conclusion regarding human intelligence follow from his focus on the details of the process as opposed to the channels that he finds so broad relative to the details. The question, he says, is "one of scale, of level of focus" (Gould 1989:289). Interestingly, Richard Dawkins, who because of the argument against divine design that he develops in *The Blind Watchmaker*, is often incorrectly cited as preaching a purely random evolutionary process, occasionally chooses to focus on the constraining, necessary aspects of evolutionary development. Most of biology, he contends, is the study of details, of the way living things actually are. But a subset of these facts is not only true but *a priori* very likely to be true. "Universal Darwinism"—the fact that organized complexity has to evolve, on Earth and elsewhere, either by divine design or by natural selection—is such a fundamental necessity. So also, he believes, are the digital nature of hereditary information, the triple nature of the genetic code, and the evolution of sex (Dawkins 1992).

A full-blown exposition of a "channel view" of biological evolution is presented by Christian de Duve in *Blueprint for a Cell* (1991) and, in a more pronounced fashion, in *Vital Dust* (1995). In the first book, de Duve explores in detail the early stages of the emergence of life, suggesting his thioester model to account for these stages. The picture that he draws is one of constrained development, in which "given a defined set of substrates, a complete assortment of rudimentary catalysts, and a reproducible supply of energy-rich thioesters, metabolism must perforce develop along lines that, though invisible, are determined by the properties of the catalysts and by the nature of their substrates" (De Duve 1991:213). This deterministic pathway, de Duve believes, could be extended to the common ancestor cell and perhaps even further. According to this view, given the same conditions as on the early Earth, life elsewhere will develop along the same chemical lines. In *Vital Dust*, de Duve reiterates even more forcefully his position on extraterrestrial life, speaking of "the living cosmos" in which "trillions of biospheres abound" (De Duve 1995:292–293). He also presents his position on the question of extraterrestrial intelligence. De Duve agrees with the evolutionary logic that leads Mayr, Gould, and others to claim the uniqueness of human intelligence. He outlines in detail the many key events involved in the evolution of the human species in which chance played a crucial role. Indeed, a slight change in any one of these events would have produced an entirely different history of life. According to de Duve, "This reasoning cannot be faulted without dismantling the solid edifice of modern evolutionary theory. . . . Should things start all over again, here or elsewhere, the final outcome would not be the same. But how different would it be?" (De Duve 1995:294).

It is de Duve's claim that "the constraints of chance," of which he gives many examples relevant to the evolution of complex forms of life, are responsible for very stringent limits on evolutionary contingency. He points out that "there is plenty of room in my reconstruction for the development of differently shaped evolutionary trees on the other planets where life has taken hold. But certain directions may carry such decisive selective advantages as to have high probability of occurrence elsewhere as well" (De Duve 1995: 296–297; Bieri 1998). This conclusion holds very strongly in the case of the development of a neuronal system, which carries much selective advantage with it. De Duve thus speculates that the drive prevalent on Earth toward larger brains, and therefore toward intelligence and the ability to communicate might also prevail on other inhabited planets. We are not alone, he concludes, and the number of biospheres that have developed intelligence is significant. Yet because of our technological limitations and the intrinsic limit set by the speed of light on the possibility of contact in space, de Duve believes that the chance of communicating with other civilizations is very slender. However, the extraordinary significance of achieving contact with even one or two alien civilizations is enough justification for the SETI endeavor (De Duve 1995:297). Many in the SETI community, pointing out the relatively low cost of the project and the enormous benefits in case of success, share the notion that the effort is worthwhile even if achievement is doubtful (Gott 1995:182). As noted by Jill Tarter, for the first time the human species has the ability to try to solve experimentally the perennial question of its status in the universe. "For those who are inclined to search, there is at least the hope of a successful conclusion in the foreseeable future" (Tarter 1995:18).

REAFFIRMING THE EVOLUTIONARY WORLDVIEW

This chapter opened with Stephen Jay Gould's declaration that he does not view the discovery of microbial fossils in a meteorite from Mars as the greatest scientific revolution since Copernicus or Darwin. It was later made clear that for Gould and other evolutionary biologists there is a distinction between the emergence of early forms of life and the evolution of more complex organisms, especially the human species and its unique intelligence. This view is not shared by many others, including scientists, philosophers, and theologians, who do not dwell much on the distinction and for whom the question of extraterrestrial life is basically the question of extraterrestrial intelligence. They believe that the discovery of life beyond Earth would have a tremendous impact on science, culture, philosophy, and religion. The physicist Paul Davies, professor of natural philosophy at the University of Adelaide, South Australia, contends that "the discovery of a single extraterrestrial microbe, if it could be shown to have evolved independently of life on Earth, would drastically alter our world view and change our society as profoundly as the Copernican and Darwinian revolutions. It could truly be described as the greatest scientific discovery of all time. In the more extreme case of the detection of an alien message, the likely effects on mankind would be awesome" (Davies 1995:xii). The reasons Davies gives for this view lead him to conclude that the discovery of life beyond Earth would give us cause to believe that we, the human species, "in our humble way, are part of a larger, majestic process of cosmic self-knowledge" (Davies 1995:129).

The fundamental reason for such significance being attached to the search for extraterrestrial life lies in the apparent need for a "meaningful universe." A universe teeming with life and thought is viewed by many as a meaningful universe, while the existence of life and consciousness on only one tiny planet as the result of an "evolutionary accident" amounts to an absurd, senseless universe. For Jacques Monod, who believed that "man is alone in the unfeeling immensity of the universe, out of which he emerged only by chance," this knowledge served as the foundation of an existentialist philosophy (Monod 1974:167). Christian de Duve, on the other hand, admits that he does not relish the thought of life on Earth being an oddity in the universe. For him, meaning is to be found in "the structure of the universe, which happens to be such as to produce thought by way of life and mind." Through the faculty of thought, the universe can reflect upon itself, discovering its structure. This, and the ability of thought to apprehend "truth, beauty, goodness and love" make the universe meaningful for de Duve (De Duve 1995:301).

An interesting twist to these ideas is found in the view expressed by the astronomer Ben Zuckerman of the University of California, Los Angeles. If the Milky Way abounds in all kinds of supercivilizations and bizarre forms of life, he says, then "life is but a commonplace extension of cosmic evolution following the Big Bang and we human beings are insignificant—mere cosmic insects." But what if this is not so? asks Zuckerman. What if we possess the most advanced brains in the galaxy? "Surely if we are alone, then we were meant to play a more noble role than that currently in evidence on our troubled globe" (Zuckerman 1995:xii). Similar sentiments have been expressed by theologians. The notion of the pivotal role of humanity in the divine scheme of things is central to Western religions, and clearly any threat to this role is bound to raise concern and to require either a non-literal interpretation of religious dogmas or a complete rejection. Donald Goldsmith suggests dividing the Judeo-Christian response to extraterrestrial life into two basic categories, "the Earth-centered and the great-glory camps" (Goldsmith 1997a:234). The Earth-centered position, echoed in the misgivings expressed by Zuckerman regarding a universe abundant with life, is mainly concerned with the disturbing possibility that if God's attention is not focused uniquely on mankind, then his purposes might be better served and advanced on another planet (see Davies 1995:50). The great-glory camp, on the other hand, regards news about possible life on Mars or elsewhere as a manifestation of God's power. It should be pointed out in passing that these controversies are not new, but rather, as fully attested by historians of ideas, have characterized the debate on the "plurality of worlds" in different periods, especially in the seventeenth and eighteenth centuries (Dick 1982; Crowe 1986; Dick 1996).

What is the deep philosophical question involved in the presence or absence of extraterrestrial life? The different evaluations of a meaningful or meaningless universe and the various religious reactions to astrobiological research are tightly bound to the question of purpose in the universe. As analyzed by Rodney L. Taylor, a professor of religious studies at the University of Colorado, the religious search for meaning is not going to be affected by the discovery of life on Mars or by any other evolutionary or astronomical discovery as long as the essence of the religious vision itself is not touched. This essence, as expressed by the philosopher W. T. Stace, is the faith that "there is a plan and purpose in the world, that the world is a moral order, that in the end all things are for the best." As long as the tenet of purpose in the world and humankind's place within it is retained, religion can incorporate into its "root metaphor" any findings of astronomy, physics, or biology (Taylor 1997).

As repeatedly pointed out in this book, the basic assumption of scientific research on evolution and its extension to the origin-of-life stage is that no supernatural, purposeful intervention is involved in the natural world. Historically, the scientific study of the origin of life became possible only after the evolutionary worldview was established. The underlying philosophical theme of current origin-of-life theories is that life emerged from a lifeless world through a continuous process governed by physicochemical mechanisms. This theme and the claim of direct or indirect purposeful interference by a supernatural designer present opposite accounts of physical reality and are thus mutually exclusive. Biologists differ in their views on the nature of evolution, disagreeing about the relative contributions of contingency and "constraints of chance" to the evolutionary process. They debate whether convergent evolution might or might not be relevant to the development of extraterrestrial intelligence. Yet it is a misinterpretation of the basic tenets of the Darwinian worldview to claim that Gould, Dawkins, and others hold that "beneath it all is simple chaos" and to conclude that "the concept of alien life is, therefore, fundamentally anti-Darwinian" (Davies 1995:76, 73). On the contrary, the search for extraterrestrial life, an extension of origin-of-life research, is fundamentally Darwinian.

Research into the origin of life on Earth, as much as the claim for the possibility of extraterrestrial life and the search for it, rests on the assumption of the evolutionary continuity between inanimate matter and life, in the context of the Copernican view of the universe. The historian Steven J. Dick, one of the major contributors to our understanding of the history of ideas about the plurality of worlds, describes the concept of abundant life in the universe as sufficiently comprehensive to qualify as a cosmological worldview, which has recently become scientifically testable. In his historical analysis he points out that not by chance did the question of extraterrestrial life become a major intellectual issue at the beginning of the seventeenth century, when the Copernican theory began to take hold, and not by chance was this idea accepted at the end of the same century. The projection of mind into space is seen by Dick as "a metaphysical completion of the scientific revolution, the final act that in the aftermath of man's greatest triumphs in understanding Nature, denied him on the universal scale the uniqueness of the only unique quality that he still possessed: mind" (Dick 1982:188). During most of history, the cosmological worldview centered on the physical universe; the concept of cosmic evolution was part of it. With the application of Darwinian ideas to the universe at large, Dick notes, we can talk now about a "biophysical cosmology." The basic postulate of this cosmology is that "if extrasolar planets did exist, chemical evolution and the origins and evolution of life had taken their course, according to Darwinian principles, on each planet" (Dick 1997:785–786).

Is this view meaningful or meaningless? Does the answer to this question depend on whether other intelligent species have evolved beyond Earth? Let me point out that from a naturalistic, evolutionary point of view, asking whether the universe is meaningful or purposeful is begging the question by assuming that purpose and meaning could belong to physical and biological reality. A "meaningless universe" is meaningless in a naturalistic vocabulary. Meaning, purpose, and values belong to the realm of human activity and thought. Even if their development might be traced back to our biological inheritance, they are nonetheless the result of the activities and mechanisms characteristic of human society. I find this view, which frees humankind from its role in the cosmological scheme of things and from its dependence on an external source of authority, to be meaningful and liberating.

Contrary to the outdated image of the scientific enterprise as a search for and collection of facts, the realization that many non-empirical factors are involved in determining scientific positions and in the adoption of scientific theories leads to the notion of theoretical and philosophical decision, or commitment. Research into the origin of life and the search for extraterrestrial life are a clear case in point, because here the weight of the philosophical commitment is much greater than in more conventional scientific fields. As long as no empirical evidence of life beyond Earth has been found, and as long as no scientific theory has succeeded in providing a fully convincing account of the emergence of life on Earth, the adoption of an evolutionary point of view toward the question of life's origin and the rejection of the idea of purposeful design involve a very strong philosophical commitment. Not only is this commitment based on the achievements of the various sciences in the last hundred years, it also proves to be a fruitful guide for research. Furthermore, growing scientific knowledge clearly strengthens the notion that the Earth, the sun, and our galaxy have no privileged status in the universe, that the human species is not unique among species regarding its origin, and that the chemistry of life as we know it is cosmically abundant. With the development of new sophisticated astronomical and biological technologies, the prospect of basing our philosophical commitment on specific empirical evidence seems more promising than ever. Both scientifically and philosophically, research into the origin of life on Earth and the search for extraterrestrial life are closely interrelated. Success in either of these areas will generally boost research and strengthen the tenets of biophysical cosmology.

Bibliography

Abel, E. L. 1973. *Ancient Views on the Origins of Life.* Rutherford, N.J.: Fairleigh Dickinson University Press.

Albee, A. 1998. "Mars Global Surveyor science update." *Planetary Report* 18, no. 2:11.

Alexander, J., and C. B. Bridges. 1926–1928. "Some physico-chemical aspects of life, mutation and evolution." In J. Alexander, ed., *Colloid Chemistry, Theoretical and Applied,* vol. 2. New York: Chemical Catalog.

Allègre, C. J., and S. H. Schneider. 1994. "The evolution of the earth." *Scientific American,* October, 44–51.

Allen, D. A., and D. T. Wickramasinghe 1987. "Discovery of organic grains in comet Wilson." *Nature* 329:615–616.

Allen, G. 1975. *Life Science in the Twentieth Century.* New York: Wiley.

Anders, E. 1996. "Evaluating the evidence for past life on Mars." *Science* 274:2119–2120.

Anderson, C. M. 1998. "Europa: layers of mystery." *Planetary Report* 18, no. 5:12–18.

Aristotle. 1952a. *On the Parts of Animals.* Trans. W. Ogle. In *The Works of Aristotle* 2:161–229. Chicago: Encyclopaedia Britannica.

———. 1952b. *On the Generation of Animals.* Trans. A. Platt. In *The Works of Aristotle* 2:255–331. Chicago: Encyclopaedia Britannica.

———. 1952c. *History of Animals.* Trans. D'Arcy W. Thompson. In *The Works of Aristotle* 2:7–158. Chicago: Encyclopaedia Britannica.

———. 1961. *Physics.* Trans. R. Hope. Lincoln: University of Nebraska Press.

———. 1991. *De Anima.* Trans. R. D. Hicks. Buffalo: Prometheus Books.

Arrhenius, G., J. L. Bada, G. F. Joyce, A. Lazcano, S. L. Miller, and L. E. Orgel. 1999. "Origin and ancestor: Separate environments." *Science* 283:792.

Arrhenius, S. 1909. *Worlds in the Making.* Trans. H. Born. London and New York: Harper.

Augustine. 1982 [415]. *The Literal Meaning of Genesis.* 2 vols. Trans. J. H. Taylor. New York: Newman.

Bachmann, P. A., P. L. Luisi, and J. Lang. 1992. "Autocatalytic self-replicating micelles as models for prebiotic structures." *Nature* 357:57–59.

Bacon, F. 1879. *The Works of Francis Bacon,* vol. 1 London: Reever.

Bada, J. L. 1995. "Origins of homochirality." *Nature* 374:594–595.

———, S. L. Miller, and M. Zhao. 1994. "Amino acid stability in hydrothermal vents." *Origins Life Evol. Biosphere* 24:364–365.

———, D. P. Glavin, G. D. McDonald, and L. Becker. 1998. "A search for endogenous amino acids in Martian meteorite ALH84001." *Science* 279:362–369.

Bailey, J., A. Chrysostomou, J. H. Hough, et al. 1998. "Circular polarization in star-formation regions: Implications for biomolecular homochirality." *Science* 281:672–674.

Balter, M. 1998. "Did life begin in hot water?" *Science* 280:31.

Baly, E. C. C., J. B. Davies, M. R. Johnson, and H. Shanassy. 1927. "The photosynthesis of naturally occurring compounds. 1—The action of ultra-violet light on carbonic acid." *Proc. Royal Soc. London* A 16:197–202.

Bar-Nun, A., E. Kochavi, and S. Bar-Nun. 1994. "An assemblage of free amino acids as a possible prebiotic enzyme." *J. Mol. Evol.* 39:116–122.

Baross, J. A., and S. E. Hoffman. 1985. "Submarine hydrothermal vents and associated gradient environments as sites for the origin and evolution of life." *Origins Life Evol. Biosphere* 15:327– 345.

Bartel, D. P., and J. W. Szostak. 1993. "Isolation of new ribozymes from a large pool of random sequences." *Science* 261:1411–1418.

Beckwith, S. V. W., and A. I. Sargent. 1996. "Circumstellar disks and the search for neighbouring planetary systems." *Nature* 383:139–144.

Behe, M. J. 1996. *Darwin's Black Box*. New York: Free Press.

Bell, J. F. 1996. "Evaluating the evidence for past life on Mars." *Science* 274:2121–2122.

Bergson, H. 1911 [1907]. *Creative Evolution*. Trans. A. Mitchell. New York: Holt. Reprinted 1983. Lanham, Md.: University Press of America.

Bernal, J. D. 1951. *The Physical Basis of Life*. London: Routledge and Kegan Paul.

———. 1965. "Discussion." In S. W. Fox, ed., *The Origin of Prebiological Systems and of their Molecular Matrices*. New York: Academic Press.

———. 1967. *The Origin of Life*. London: Weidenfeld and Nicolson.

Bernstein, M. P., S. C. Stanford, L. J. Allamandola, J. S. Gillette, S. J. Clemett, and R. N. Zare. 1999. "UV irradiation of polycyclic aromatic hydrocarbons in ices: Production of alcohols, quinones, and ethers." *Science* 283: 1135–1138.

Biebricher, C. K. 1983. "Darwinian selection of RNA molecules *in vitro*." In M. K. Hecht, B. Wallace, and G. T. Prance, eds., *Evolutionary Biology*, vol. 16. New York: Plenum.

———, M. Eigen, and R. Luce. 1981. "Product analysis of RNA generated *de novo* by Qß replicase." *J. Mol. Biol.* 148:369–390.

Bieri, R. 1998. "Huminoids on other planets?" In M. Ruse, ed., *Philosophy of Biology*, 272–278. Amherst: Prometheus Books.

Biot, J.-B. 1844. "Communication d'une note de M. Mitscherlich." *Comptes rendus de l'Acad. des sci.* 19:720.

Black, D. C. 1995. "Completing the Copernican revolution: The search for other planetary systems." In G. Burbidge, and A. Sandage, eds., *Annual Review of Astronomy and Astrophysics* 33:359.

Böhler, C., P. E. Nielsen, and L. E. Orgel. 1995. "Template switching between PNA and RNA oligonucleotides." *Nature* 376:578–581.

Bonner, W. A. 1991. "The origin and amplification of biomolecular chirality." *Origins Life Evol. Biosphere* 21:59–111.

Borowska, Z. K., and D. C. Mauzerall. 1987. "Efficient near-ultraviolet-light-induced formation of hydrogen by ferrous hydroxide." *Origins Life Evol. Biosphere* 17:251–259.

Bowler, P. J. 1984. *Evolution—the History of an Idea*. Berkeley: University of California Press.

Brack, A. 1993. "Liquid water and the origin of life." *Origins Life Evol. Biosphere* 23:3–10.

Bradley, D. 1994. "A new twist in the tale of nature's asymmetry." *Science* 264:908.

Bradley, J. P., R. P. Harvey, and H. Y. McSween, Jr. 1997. "No 'nanofossils' in Martian meteorite." *Nature* 390:454–455.

Bradley, W. L., and C. B. Thaxton. 1994. "Information & the origin of life." In J. P.

Moreland, ed., *The Creation Hypothesis: Scientific Evidence for an Intelligent Designer*, 173–210. Illinois: InterVarsity Press.

Brooke, J. H. 1991. *Science and Religion*. Cambridge: Cambridge University Press.

Buffon, G.-L. 1749. *Histoire naturelle, générale et particulière*, vol 2. Paris.

———. 1962 [1778]. *Les époques de la nature*. Ed. J. Roger. Paris: Musée d'histoire naturelle.

Burnet, J. 1930. *Early Greek Philosophy*. 4th ed. London: A. and C. Black.

Butler, P. R., and G. W. Marcy. 1997. "The Lick Observatory planet search." In C. B. Cosmovici, S. Bowyer, and D. Werthimer, eds., *Astronomical and Biochemical Origins and the Search for Extraterrestrial Life in the Universe*, 331–342. Proceedings of the 5th International Conference on Bioastronomy. Bologna, Italy: Editrice Compositori.

Butt, C. J., C. R. Woese, J. C. Venter et al. 1996. "Complete genome sequence of the methanogenic archaeon, Methanococcus jannaschii." *Science* 273:1058–1073.

Cairns-Smith, A. G. 1985. *Seven Clues to the Origin of Life*. Cambridge, U.K.: Cambridge University Press.

———, A. J. Hall, and M. J. Russell. 1992. "Mineral theories of the origin of life and an iron sulfide example." *Origins Life Evol. Biosphere* 22:161–180.

Cech, T. R., and O. C. Uhlenbeck. 1994. "Hammerhead nailed down." *Nature* 372:39–40.

Chakrabarti A. C., and D. W. Deamer. 1994. "Permeation of membranes by the neutral form of amino acids and peptides: Relevance to the origin of peptide translocation." *J. Mol. Evol.* 39, no. 1.1–5.

Chyba, C. F. 1997a. "Life on other moons." *Nature* 385:201.

———. 1997b. "Catastrophic impacts and the Drake equation." In C. B. Cosmovici, S. Bowyer, and D. Werthimer, eds., *Astronomical and Biochemical Origins and the Search for Life in the Universe*, 157–164. Proceedings of the 5th International Conference on Bioastronomy. Bologna, Italy: Editrice Compositori.

———. 1998. "The stuff of life: Why water?" *Planetary Report* 18, no. 3:16–17.

———, and C. Sagan. 1992. "Endogenous production, exogenous delivery and impact shock synthesis of organic molecules: An inventory for the origins of life." *Nature* 355:125–132.

Clarke, D. W., and J. P. Ferris. 1997. "Chemical evolution on Titan: Comparisons to the prebiotic Earth." *Origins Life Evol. Biosphere* 27:225–248.

Clemett, S. J., and R. N. Zare. 1996. "Evaluating the evidence for past life on Mars. Response." *Science* 274:2122–2123.

Cocconi, G., and P. Morrison. 1959. "Searching for interstellar communications." *Nature* 184:844–846.

Cohen, J. 1995a. "Getting all turned around over the origins of life on Earth." *Science* 267:1265–1266.

———. 1995b. "Novel center seeks to add spark to origins of life." *Science* 270:1925–1926.

Coleman, W. 1965. "Cell, nucleus and inheritance: A historical study." *Proc. Amer. Phil. Soc.* 109:124–158.

Cooper, J. M. 1987. "Hypothetical necessity and natural teleology." In A. Gotthelf and J. G. Lennox, eds., *Philosophical Issues in Aristotle's Biology*, 243–274. Cambridge: Cambridge University Press.

Corliss, J. B. 1990. "Hot springs and the origin of life." *Nature* 347:624.

———, J. Dymond, L. I. Gordon, J. M. Edmond, R. P. von Herzen, R. D. Ballard, K. Green, D. Williams, A. Bainbridge, K. Crane, and T. H. van Andel. 1979. "Submarine thermal springs on the Galápagos Rift." *Science* 203:1073–1083.

Cornford, F. M. 1957. *Plato's Cosmology*. Translation of Plato's *Timaeus*. New York: Library of Liberal Arts.

Crick, F. 1981. *Life Itself*. New York: Simon and Schuster.

———. 1990. *What Mad Pursuit*. New York: Penguin Books.

Crick, F., and L. E. Orgel. 1973. "Directed panspermia." *Icarus* 19:341–346.

Crick, F. H. C. 1968. "The origin of the genetic code." *J. Mol. Biol.* 38:367–379.

Cronin, J. R., and S. Chang. 1993. "Organic matter in meteorites: Molecular and isotopic analyses of the Murchison meteorite." In J. M. Greenberg et al., eds., *The Chemistry of Life's Origin*, 205–258. Dordrecht: Kluwer Academic Publishers.

Cronin, J. R., and S. Pizzarello. 1997. "Enantiomeric excess in meteoritic amino acids." *Science* 275:951–955.

Crowe, M. J. 1986. *The Extraterrestrial Life Debate 1750–1900: The Idea of a Plurality of Worlds from Kant to Lowell*. New York and Cambridge: Cambridge University Press.

Crutchfield, J. P., D. J. Farmer, N. H. Packard, and R. S. Shaw. 1986. "Chaos." *Scientific American*, December, 38–49.

Cunningham, G. E. 1998. "Mars Global Surveyor: The saga of the solar array." *Planetary Report* 18, no. 2:4–10.

Cuvier, G. 1863 [1817]. *The Animal Kingdom Arranged after Its Organization*. 4 vols. Millwood, N.Y.: Kraus Reprints.

Darwin, C. 1963 [1859]. *The Origin of Species*. New York: Washington Square Press.

Darwin, E. 1804. *The Temple of Nature*. Reprinted Elmsford, N.Y.: Pergamon.

Darwin, F. 1969 . *The Life and Letters of Charles Darwin*. 3 vols. New York: Johnson Reprint Corp. (Originally published in 1887 by Murray, London.)

———, F. 1995. *The Life of Charles Darwin*. London: Senate. (Originally published in 1902 by Murray, London).

———, and A. C. Seward. 1903. *More Letters of Charles Darwin*. 2 vols. London: Murray.

Davies, P. 1995. *Are We Alone?* New York: Basic Books.

Dawkins, R. 1978. *The Selfish Gene*. New York: Oxford University Press.

———. 1986. *The Blind Watchmaker*. London: Penguin Books.

———. 1992. "Universal biology." *Nature* 360:25–26.

———. 1997. "Obscurantism to the rescue." *Quarterly Review of Biology* 72, no. 4:397–399.

Deamer, D. W. 1997. "The first living systems: A bioenergetic perspective." *Microbiology and Molecular Biology Review* 61:2, 239–261.

———, and R. M. Pashley. 1989. "Amphiphilic components of carbonaceous meteorites." *Origins Life Evol. Biosphere* 19:21–38.

De Duve, C 1991 *Blueprint for a Cell*. Burlington, N.C.. Neil Patterson Publishers.

———. 1995. *Vital Dust*. New York: Basic Books.

———, and S. L. Miller. 1991. "Two-dimensional life?" *Proc. Natl. Acad. Sci. USA* 88:10014–10017.

Descartes, R. 1958 [1641]. *Meditations of First Philosophy*. In *Philosophical Writings*, trans. N. K. Smith. New York: Modern Library.

———. 1965 [1637]. *Discourse on Method*. Trans. P. Olscamp. Indianapolis: Bobbs-Merrill.

———. 1983 [1644]. *Principles of Philosophy*. Trans. V. Miller and R. Miller. Dordrecht: Reidel.

D'Holbach, P. H. 1970 [1770]. *The System of Nature: or Laws of the Moral and Physical World*. Trans. H. D. Robinson. New York: Burt Franklin.

Dick, S. J. 1982. *Plurality of Worlds: The Origins of the Extraterrestrial Life Debate from Democritus to Kant*. New York and Cambridge: Cambridge University Press.

———. 1996. *The Biological Universe: The Twentieth Century Extraterrestrial Life Debate and the Limits of Science*. New York and Cambridge: Cambridge University Press.

———. 1997. "The biophysical cosmology: The place of bioastronomy in the history of science." In C. B. Cosmovici, S. Bowyer, and D. Werthimer, eds., *Astronomical and Biochemical Origins and the Search for Life in the Universe*, 785–788. Proceedings of the 5th International Conference on Bioastronomy. Bologna, Italy: Editrice Compositori.

Diderot, D. 1966 [1769]. *D'Alembert's Dream and Rameau's Nephew*. Trans. L. W. Tancock. Harmondsworth, U.K.: Penguin Books.

Digregorio, B. E., with G. V. Levin, and P A Straat. 1997. *Mars, the Living Planet*. Berkeley: Frog.

Diodorus Siculus. 1960. *History*. Cambridge: Harvard University Press. (Original text from first century B.C.E.)

Dobzhansky, T. 1973. "Nothing in biology makes sense except in the light of evolution." *American Biology Teacher* 35:125–129.

Domingo, E., D. Sabo, T. Tanaguchi, and C. Weissmann. 1978. "Nucleotide sequence heterogeneity of an RNA phage population." *Cell* 13:735.

Doolittle, R. F. 1998. "Microbial genomes opened up." *Nature* 392:339–342.

Dover, G. A. 1993. "On the edge." *Nature* 365:704–706.

Drake, F., and D. Sobel. 1992. *Is Anyone Out There?* New York: Delacorte Press.

Driesch, H. 1914. *The Problem of Individuality*. London: Macmillan.

Dubos, R. 1950. *Louis Pasteur: Free Lance of Science*. Boston: Little, Brown.

Dyson, F. 1982. "A model for the origin of life." *J. Mol. Evol.* 18:344–350.

———. 1985. *Origins of Life*. Cambridge: Cambridge University Press.

Easterbrook, G. 1997. "Science and God: A warming trend?" *Science* 277:890–893.

Ehrenfreund, P. 1999. "Molecules on a space odyssey." *Science* 283:1123–1124.

Eigen, M. 1971. "Self-organization of matter and the evolution of biological macromolecules." *Naturwissenschaften* 58:465–523.

———. 1992. *Steps towards Life*. Oxford: Oxford University Press.

———. 1993. "Viral quasispecies." *Scientific American*, July, 42–49.

———, and P. Schuster. 1977. The hypercycle, Part A: The emergence of the hypercycle." *Naturwissenschaften* 64:541–565.

———, and P. Schuster. 1978. "The hypercycle, Part C: The realistic hypercycle." *Naturwissenschaften* 65:341–369.

———, W. Gardiner, P. Schuster, and R. Winkler-Oswatitsch. 1981. "The origin of genetic information." *Scientific American*, April, 78–118.

Ellington, A. D., M. P. Robertson, and J. Bull. 1997. "Ribozymes in wonderland." *Science* 276:546–547.

Emmeche, C. 1994. *The Garden in the Machine*. Princeton, N.J.: Princeton University Press.

Endler, J. A. 1986. "The newer synthesis? Some conceptual problems in evolutionary biology." In R. Dawkins and M. Riddley, eds., *Oxford Surveys in Evolutionary Biology* 3:224–243. Oxford: Oxford University Press.

Ertem, G., and J. P. Ferris. 1996. "Synthesis of RNA oligomers on heterogeneous templates." *Nature* 379:238–240.

Eschenmoser, A. 1994. "Chemistry of potentially prebiological natural products." *Origins Life Evol. Biosphere* 24:389–423.

Farley, J. 1977. *The Spontaneous Generation Controversy from Descartes to Oparin*. Baltimore: Johns Hopkins University Press.

———, and G. L. Geison. 1974. "Science, politics and spontaneous generation in nineteenth-century France: The Pasteur-Pouchet debate." *Bulletin of the History of Medicine* 48, no. 2:161–198.

Ferris, J. P. 1987. "Prebiotic synthesis: Problems and challenges." *Cold Spring Harbor Symp. Quant. Biol.* 55:29–35.

———. 1994. "Chemical replication." *Nature* 369:184–185.

———, Kamaluddin, and G. Ertem. 1990. "Oligomerization reactions of oligonucleotides on montmorillonite clay." *Origins Life Evol. Biosphere* 20:279–291.

———, and G. Ertem. 1992. Oligomerization of ribonucleotides on montmorillonite: Reactions of the 5'-phosphoimidazolide of adenosine." *Science* 257:1387–1389.

———, A. R. Hill, Jr., R. Liu, and L. E. Orgel. 1996. "Synthesis of long prebiotic oligomers on mineral surfaces." *Nature* 381:59–61.

Fieser, L. F., and M. Fieser. 1961. *Organic Chemistry*. 3rd ed. New York: Reinhold Publishing.

Fleischmann, R. D., M. D. Adams, O. White, R. A. Clayton, E. F. Kirkness, A. R. Kerlavage, C. J. Bult, J. F. Tomb, B. A. Dougherty, J. M. Merrick, et al. 1995. "Whole-genome random sequencing and assembly of Haemophilus influenzae Rd." *Science* 269:496–512.

Folk, R. L. 1997. "Nannobacteria: Size limits and evidence. Response." *Science* 276:1777.

Forterre, P., F. Confalonieri, F. Charbonnier, and M. Duguet. 1995. "Speculations on the origin of life and thermophily: Review of available information on reverse gyrase suggests that hyperthermophilic procaryotes are not so primitive." *Origins Life Evol. Biosphere* 25:235–249.

Fox, S. W. 1980. "Life from an orderly cosmos." *Naturwissenschaften* 67:576–581.

———. 1984. "Proteinoid experiments and evolutionary theory." In M. W. Ho, and P. T. Saunders, eds., *Beyond Neo-Darwinism*, 15–60. New York: Academic Press.

———, and K. Harada. 1958. "Thermal copolymerization of amino acids to a product resembling protein." *Science* 128:1214.

———, and C. R. Windsor. 1970. "Synthesis of amino acids by the heating of formaldehyde and ammonia." *Science* 170:984–986.

———, and K. Dose. 1977. *Molecular Evolution and the Origin of Life.* New York: Marcel Dekker.

Fruton, J. S. 1972. *Molecules and Life.* New York: John Wiley.

Fry, I. 1995. "Are the different hypotheses on the emergence of life as different as they seem?" *Biology & Philosophy* 10:389–417.

———. 1996. "On the biological significance of the properties of matter: L. J. Henderson's theory of the fitness of the environment." *Journal of the History of Biology* 29:155–196.

Gaffey, M. J. 1997. "The early solar system." *Origins Life Evol. Biosphere* 27:185–203.

Galtier, N., N. Tourasse, and M. Gouy. 1999. "A nonhyperthermophilic common ancestor to extant life forms." *Science* 283:220–221.

Garrison, W. M., D. C. Morrison, J. G. Hamilton, A. A. Benson, and M. Calvin. 1951. "Reduction of carbon dioxide in aqueous solutions by ionizing radiation." *Science* 114:416–418.

Geison, G. L. 1969. "The protoplasmic theory of life and the vitalist-mechanist debate." *Isis* 60:273–292.

———. 1995. *The Private Science of Louis Pasteur.* Princeton, N.J.: Princeton University Press.

Gibor, A, ed., 1976. *Conditions for Life: Readings from* Scientific American. San Francisco: W. H. Freeman.

Gibson, E. K., Jr., D. S. McKay, and K. L. Thomas-Keprta. 1998. "The case for life on Mars. Part 2: Data support the hypothesis." *Bioastronomy News* 10, no. 3:1–6.

Gilbert, W. 1986. "The RNA world." *Nature* 319:618.

Glanz, J. 1997. "Worlds around other stars shake planet birth theory." *Science* 276:1336–1339.

Gleick, J. 1987. *Chaos: Making a New Science.* New York: Viking.

Gogarten-Boekels, M., E. Hilario, and J. P. Gogarten. 1995. "The effects of heavy meteorite bombardment on the early evolution—the emergence of the three domains of life." *Origins Life Evol. Biosphere* 25:251–264.

Goldanskii, V. L., and V. V. Kuzmin. 1989. "Spontaneous breaking of mirror symmetry in nature and the origin of life." *Sov. Phys. Usp.* 32:1–29.

Goldsmith, D. 1997a. *The Hunt for Life on Mars.* New York: Dutton.

———. 1997b. *Worlds Unnumbered: The Search for Extraterrestrial Planets.* Sausalito, Calif.: University Science Books.

———. 1999. "New telescope will turn a keen ear on E.T." *Science* 283:914.

Goodwin, B., and P. Saunders, eds. 1989. *Theoretical Biology: Epigenetic and Evolutionary Order from Complex Systems.* Edinburgh: Edinburgh University Press.

Gott, R. J. III. 1993. "Implications of the Copernican principle for our future prospects." *Nature* 363:315–319.

Gott, R. J. III. 1995. "Cosmological SETI frequency standards." In B. Zuckerman, and M. H. Hart, eds., *Extraterrestrials: Where Are They?* 2nd ed., 173–183. Cambridge, U.K.: Cambridge University Press.

Gotthelf, A. 1987. "Aristotle's conception of final causality." In A. Gotthelf, and J. G. Lennox, eds., *Philosophical Issues in Aristotle's Biology*, 199–203. Cambridge: Cambridge University Press.

Gould, S. J. 1977. "On heroes and fools in science." In S. J. Gould, *Ever Since Darwin*, 201–206. New York: W. W. Norton.

———. 1982. "Bathybius and eozoon." In S. J. Gould, *The Panda's Thumb*, 236–244. New York: W. W. Norton.

———. 1989. *Wonderful Life*. New York: W. W. Norton.

———. 1992. "Impeaching a self-appointed judge." *Scientific American*, July, 118–121.

———. 1997. "Nonoverlapping Magisteria." *Natural History* 106, no. 2:16–22, 60–62.

———, and R. C. Lewontin. 1979. "The spandrels of San Marco and the Panglossian paradigm: A critique of the adaptationist programme." *Proc. Royal Soc. London.* B 205:581–598.

Graham, L. R. 1987. *Science, Philosophy and Human Behavior in the Soviet Union.* New York: Columbia University Press.

Greenberg, J. M. 1995. "Chirality in interstellar dust and in comets: Life from dead stars." In D.B. Cline, ed., *Physical Origin of Homochirality in Life*, 185–210. Proceedings from the Symposium in Santa Monica. Santa Monica: AIP Press.

———, and C. X. Mendoza-Gómez. 1993. "Interstellar dust evolution: A reservoir of prebiotic molecules." In J. M. Greenberg et al., eds., *The Chemistry of Life's Origins*, 1–32. Dordrecht: Kluwer Academic Press.

Gray, M. W. 1996. "The third form of life." *Nature* 383:300.

Gribbin, J. 1993. *In the Beginning*. Boston: Little, Brown.

Guerrier-Takada, C., K. Gardiner, T. Marsh, N. Pace, and S. Altman. 1983. "The RNA moiety of ribonuclease P is the catalytic subunit of the enzyme." *Cell* 35:849–857.

Haeckel, E. 1866. *Generelle Morphologie der Organismen*. 2 vols. Berlin.

———. 1902. *The Riddle of the Universe*. Trans. J. McCabe. New York and London: Harper & Brothers Publishers.

Haldane, J. B. S. 1954. "The origins of life." *New Biology* 16:21–26.

———. 1967. "The origin of Life." In J. D. Bernal. *The Origin of Life*, 242–249. London: Weidenfeld and Nicolson. (Originally published in 1929 in *Rationalist Annual*, 3–10.)

Hart, M. H. 1978. "The evolution of the atmosphere of the Earth." *Icarus* 33:23–29.

Harvey, R. P., and H. Y. McSween, Jr. 1996. "A possible high-temperature origin for the carbonates in the Martian meteorite ALH84001." *Nature* 382:49–51.

Harvey, W. 1847. *On the Generation of Animals*. In *The Works of William Harvey*. London: Sydenham Society.

Hayes, J. M. 1996. "The earliest memories of life on Earth." *Nature* 384:21–22.

Hegstrom, R. A., and D. K. Kondepudi. 1990. "The handedness of the universe." *Scientific American*, January, 98–105.

Heidmann, J. 1997. *Extraterrestrial Intelligence*. Cambridge: Cambridge University Press.

Heinen, W., and A. M. Lauwers. 1996. "Organic sulfur compounds resulting from the interaction of iron sulfide, hydrogen sulfide and carbon dioxide in an anaerobic aqueous environment." *Origins Life Evol. Biosphere* 26:131–150.

Henderson, L. J. 1970 (1913). *The Fitness of the Environment*. Gloucester, Mass.: Peter Smith.

Holden, C. 1999. "Martian life: Another round." *Science* 283:1841.

Holland, W. S., J. B. Greaves, B. Zuckerman, I. Robson, et al. 1998. "Submillimetre images of dusty debris around nearby stars." *Nature* 392:788–791.

Holm, N. G. 1992. "Why are hydrothermal systems proposed as plausible environments for the origin of life?" *Origins Life Evol. Biosphere* 22:5–14

———. 1996. "Serpentinization of oceanic crust and Fischer-Tropsch type synthesis of organic compounds." *Origins Life Evol. Biosphere* 26:205–206.

———, A. G. Cairns-Smith, et al. 1992. "Future research." *Origins Life Evol. Biosphere* 22:181–190.

Hoppe, P., R. Strebel, P. Eberhardts, S. Amari, and R. S. Lewis. 1996. "Type II supernova matter in silicon carbide grain from the Murchison meteorite." *Science* 272:1314–1317.

Horgan, J. 1991. "In the beginning . . ." *Scientific American*, February, 114–125.

Hoyle, F., and C. Wickramasinghe. 1979a. *Lifecloud*. New York: Harper and Row.

———, and C. Wickramasinghe. 1979b. *Diseases from Space*. New York: Harper and Row.

————, and C. Wickramasinghe. 1981. *Evolution from Space*. London: J. M. Dent and Sons.

————, and C. Wickramasinghe. 1988. *Cosmic Life-Force*. London: J. M. Dent and Sons.

————, and C. Wickramasinghe. 1993. *Our Place in the Cosmos*. London: J. M. Dent and Sons.

Huber, C., and G. Wächtershäuser. 1997. "Activated acetic acid by carbon fixation on (Fe, Ni)S under primordial conditions." *Science* 276:245–247.

————, and ————. 1998. "Peptides by activation of amino acids with CO on (Ni, Fe)S surfaces: Implications for the origin of life." *Science* 281:670–672.

Huebner, W. F. 1987. "First polymer in space identified in comet Halley." *Science* 237:628–630.

Hull, D. 1991. "The God of the Galápagos." *Nature* 352:485–486.

Hume, D. 1966 [1799]. *Dialogues concerning Natural Religion*. New York: Hafner.

Huxley, T. H. 1869. "On the physical basis of life." *Fortnightly Review* 5:129–145.

Inoue, T., and L. E. Orgel. 1983. "A nonenzymatic RNA polymerase model." *Science* 219:859–862.

Irion, R. 1998. "Did twisty starlight set stage for life?" *Science* 281:626–627.

Irvine, W. M. 1998. "Extraterrestrial organic matter. A review." *Origin Life Evol. Biosphere* 28:365–383.

Jacob, F. 1982. *The Logic of Life*. Trans. Betty E. Spillmann. New York: Pantheon Books.

Jakosky, B. 1997. "The case for life on Mars." *Planetary Report* 17, no. 1:12–17.

————. 1998. *The Search for Life on Other Planets*. Cambridge: Cambridge University Press.

James, K. D., and A. D. Ellington. 1995. "A search for missing links between self-replicating nucleic acids and the RNA world." *Origins Life Evol. Biosphere* 25:515–530.

John Paul II. 1997. "Message to the Pontifical Academy of Sciences." *Quarterly Review of Biology* 72, no. 4:381–383.

Johnson, P. E. 1993. *Darwin on Trial*. Illinois: InterVarsity Press.

————. 1995. *Reason in the Balance*. Illinois: InterVarsity Press.

Joyce, G. F. 1989. "RNA evolution and the origins of life." *Nature* 338:217–224.

————. 1992. "Directed molecular evolution." *Scientific American*, December, 90–97.

————, G. M. Visser, C. A. A. van Boeckel, J. H. van Boom, L. E. Orgel, and J. van Westrenen. 1984. "Chiral selection in poly(C)-directed synthesis of oligo(G)." *Nature* 310:602–604.

————, and L. E. Orgel. 1993. "Prospects for understanding the origin of the RNA world." In R. F. Gesteland, and J. F. Atkins, eds., *The RNA World*, 1–25. Plainview, N.Y.: Cold Spring Harbor Laboratory Press. [See Gesteland, R. F., T. Cech, and J. F. Atkins, eds., 1998. *The RNA World*. 2nd ed. Plainview, N.Y.: Cold Spring Harbor Laboratory Press.]

Jull, A. J. T., C. Courtney, D. A. Jeffrey, and J. W. Beck. 1998. "Isotopic evidence for a terrestrial source of organic compounds found in Martian meteorites Allan Hills 84001 and Elephant Moraine 79001." *Science* 279:366–369.

Kajander, E. O., I. Kuronen, N. Ciftcioglu. 1996. "Fetal bovine serum: Discovery of nanobacteria." *Molecular Biology of the Cell* 7:3007–3007, suppl. S.

Kamminga, H. 1982. "Life from space—a history of panspermia." *Vistas in Astronomy* 26:67–86.

————. 1986. "The protoplasm and the gene." In A. G. Cairns-Smith and H. Hartman, eds., *Clay Minerals and the Origin of Life*, 1–10. Cambridge: Cambridge University Press.

————. 1988. "Historical perspective: The problem of the origin of life in the context of developments in biology." *Origins Life Evol. Biosphere* 18:1–11.

————. 1991. "The origin of life on Earth: Theory, history and method." *Uroborus* 1, no. 1: 95–110.

Kanavarioti, A. 1994. "Template-directed chemistry and the origins of the RNA world." *Origins Life Evol. Biosphere* 24:479–494.

Kant, I. 1965 [1781]. *Critique of Pure Reason*. Trans. N. Kemp Smith. New York: St. Martin's Press.

——. 1987 [1790]. *Critique of Judgment. Part 2. Critique of Teleological Judgment*. Trans. W. S. Pluhar. Indianapolis: Hackett Publishing Company.

Kaschke, M., M. J. Russell, and W. J. Cole. 1994. "[FeS/FeS$_2$]. A redox system for the origin of life." *Origins Life Evol. Biosphere* 24:43–56.

Kasting, J. F. 1993. "Earth's early atmosphere." *Science* 259: 920–926.

——. 1997a. "Habitable zones around low mass stars and the search for extraterrestrial life." *Origins Life Evol. Biosphere* 27:291–307.

——. 1997b. "Warming early Earth and Mars." *Science* 276:1213–1215.

——, D. P. Whitmire, and R. T. Reynolds. 1993. "Habitable zones around main sequence stars." *Icarus* 101:108.

——, and L. L. Brown. 1996. "Methane concentrations in the Earth's prebiotic atmosphere." *Origins Life Evol. Biosphere* 26:219–220.

Kauffman, S. A. 1991. "Antichaos and adaptation." *Scientific American*, August, 78–84.

——. 1993. *The Origins of Order: Self-Organization and Selection in Evolution*. New York: Oxford University Press.

Kenyon, D. H., and G. Steinman. 1969. *Biological Predestination*. New York: McGraw-Hill.

Kerr, R. A. 1997a. "Martian 'microbes' cover their tracks." *Science* 276:30–31.

——. 1997b. "Putative Martian microbes called microscopy artifacts." *Science* 278:1706–1707.

——. 1997c. "An ocean emerges on Europa." *Science* 276:355.

——. 1998a. "Requiem for life on Mars? Support for microbes fades." *Science* 282:1398–1400.

——. 1998b. "Geologists take a trip to the red planet." *Science* 282:1807–1809.

Kimura, M. 1983. *The Neutral Theory of Molecular Evolution*. Cambridge: Cambridge University Press.

Kirk, G. S., and J. E. Raven. 1957. *The Presocratic Philosophers*. Cambridge: Cambridge University Press.

Kitcher, P. 1982. *Abusing Science*. Cambridge, Mass.: The MIT Press.

Kochavi, E., A. Bar-Nun, and G. Fleminger. 1997. "Substrate-directed formation of small biocatalysts under prebiotic conditions." *J. Mol. Evol.* 45:342–351.

Kohler, E. R. 1973. "The enzyme theory and the origin of biochemistry." *Isis* 64:181–196.

Kok, R. A., J. A. Taylor, and W. L. Bradley. 1988. "Statistical examination of self-ordering of amino acids in proteins." *Origin Life Evol. Biosphere* 18:135–142.

Kolb, V. M., J. P. Dworkin, and S. L. Miller. 1994. "Urazole is a potential precursor to uracil." *Origins Life Evol. Biosphere* 24:107–108.

Kondepudi, D. K., and G. W. Nelson. 1985. "Weak neutral currents and the origin of biomolecular chirality." *Nature* 314:438–441.

Kruger, K., P. J. Grabowski, A. J. Zaug, J. Sands, D. E. Gottschling, and T. R. Cech. 1982. "Self-splicing RNA: Autoexcision and autocyclization of the ribosomal RNA intervening sequence of *Tetrahymena*." *Cell* 31:147–157.

Küppers, B.-O. 1990. *Information and the Origin of Life*. Cambridge, Mass.: MIT Press.

Lahav, N., and S. Nir. 1997. "Emergence of template-and-sequence-directed (TSD) synthesis: 1. A bio-geochemical model." *Origins Life Evol. Biosphere* 27:377–395.

Lamarck, J.-B. 1963 [1809]. *Zoological Philosophy*. Trans. H. Elliot. New York: Hafner.

Lambridis, H. 1976. *Empedocles: A Philosophical Investigation*. University, Ala.: University of Alabama Press.

Latour, B. 1988. *The Pasteurization of France*. Cambridge, Mass.: Harvard University Press.

Lazcano, A., and S. L. Miller. 1994. "How long did it take for life to begin and evolve to cyanobacteria?" *J. Mol. Evol.* 39:546–554.

Lee, D. H., J. R. Cranja, J. A. Martinez, K. Severin, and M. R. Ghadiri. 1996. "A self-replicating peptide." *Nature* 382:525–528.

——, K. Severin, Y. Yokobayashi, and M. R. Ghadiri. 1997. "Emergence of symbiosis in peptide self-replication through a hypercyclic network." *Nature* 390:591–594.

Léger, A., M. Pirre, and F. Marceau. 1993. "Search for primitive life on a distant planet: Relevance of O_2 and O_3 detection." *Astro. Astrophys.* 277:309–313.

Lehninger, A. L. 1970. *Biochemistry*. New York: Worth Publishers.

Lennox, J. G. 1982. "Teleology, chance and Aristotle's theory of spontaneous generation." *Journal of the History of Philosophy* 20:219–238.

Lenoir, T. 1982. *The Strategy of Life: Teleology and Mechanics in Nineteenth Century German Biology*. Dordrecht, Netherlands: Reidel.

Levine, J. S, ed. 1985. *The Photochemistry of the Atmospheres*. New York: Academic Press.

Levy, M., and S. L. Miller. 1998. "The stability of the RNA bases: Implications for the origin of life." *Proc. Natl. Acad. Sci. USA* 95:7933–7937.

Li, T., and K. C. Nicolauo. 1994. "Chemical self-replication of palindromic duplex DNA." *Nature* 369:218–221.

Lifson, S. 1997. "On the crucial stages in the origin of animate matter." *J. Mol. Evol.* 44:1–8.

Lin, D. N. C., P. Bodenheimer, and D. C. Richardson. 1996. "Orbital migration of the planetary companion of 51 Pegasi to its present location." *Nature* 380:606.

Lissauer, J. J. 1997. "Formation, frequency and spacing of habitable planets." In C. B. Cosmovici, S. Bowyer, and D. Werthimer, eds., *Astronomical and Biochemical Origins and the Search for Extraterrestrial Life in the Universe*, 289–297. Proceedings of the 5th International Conference on Bioastronomy. Bologna, Italy: Editrice Compositori.

Loeb, J. 1964 [1912]. *The Mechanistic Conception of Life*. Cambridge, Mass.: Harvard University Press.

Lovelock, J. 1979. *Gaia: A New Look at Life on Earth*. New York: Oxford University Press.

Lowes, L. L. 1998. "Ice, water, and fire: The Galileo Europa mission." *Planetary Report* 18, no. 1:12–15.

Lucretius. 1956. *De Rerum Natura*. Trans. A. D. Winspear. New York: S. A. Russell. The Harbor Press. (Original text from first century B.C.E.)

Maniloff, J. 1997. "Nannobacteria: Size limits and evidence." *Science* 276:1776.

Marcy, G. W., P. R. Butler, S. S. Vogt, D. Fischer, and J J. Lissauer. 1998. "A planetary companion to a nearby M4 Dwarf. Gliese 876." *Astrophysical Journal* 505:L147–L149.

Marcy and Butler Web site: *www.physics.sfsu.edu/~gmarcy/planetsearch/planetsearch.html*.

Margulis, L. 1981. *Symbiosis in Cell Evolution*. San Francisco: Freeman.

Marino, L. 1997. "Brain-behavior relations in primates and cetaceans: Implications for the ubiquity of factors leading to the evolution of complex intelligence." In C. B. Cosmovici, S. Bowyer, and D. Werthimer, eds., *Astronomical and Biochemical Origins and the Search for Life in the Universe*, 553–560. Proceedings of the 5th International Conference on Bioastronomy. Bologna, Italy: Editrice Compositori.

Mariotti, J. M., A. Léger, B. Mennesson, and M. Ollivier. 1997. "Detection and characterization of Earth-like planets." In C. B. Cosmovici, S. Bowyer, and D. Werthimer, eds., *Astronomical and Biochemical Origins and the Search for Extraterrestrial Life in the Universe*, 299–311. Proceedings of the 5th International Conference on Bioastronomy. Bologna, Italy: Editrice Compositori.

Mars Global Surveyor Web site: *http://mars.jpl.nasa.gov/mgs*

Maupertuis, P. L. M. 1966 [1745]. *The Earthly Venus*. Trans. S. B. Boas. New York: Johnson Reprint Corporation.

Maurette, M. 1998. "Carbonaceous micrometeorites and the origin of life." *Origins Life Evol. Biosphere* 28:385–412.

Maynard Smith, J. 1986. *The Problems of Biology*. Oxford: Oxford University Press.

———. 1995. "Life at the edge of chaos?" *New York Review of Books*, March, 2:28–30.

———, and E. Szathmáry. 1995. *The Major Transitions in Evolution*. Oxford: W. H. Freeman.

Mayor, M., and D. Queloz. 1995. "Jupiter-mass companion to a solar-type star." *Nature* 378:355–359.

———, D. Queloz, S. Udry, and J-L. Halbwachs. 1997. "From brown dwarfs to planets." In C. B. Cosmovici, S. Bowyer, and D. Werthimer, eds., *Astronomical and Biochemical Origins and the Search for Extraterrestrial Life in the Universe*, 313–330. Proceedings of the 5th International Conference on Bioastronomy. Bologna, Italy: Editrice Compositori.

Mayr, E. 1982. *The Growth of Biological Thought*. Cambridge, Mass.: Harvard University Press.

———. 1995. "The search for extraterrestrial intelligence." In B. Zuckerman, and M. H. Hart, eds., *Extraterrestrials: Where Are They?* 2nd ed. 152–156. Cambridge: Cambridge University Press.

———. 1998. "The probability of extraterrestrial intelligent life." In M. Ruse, ed., *Philosophy of Biology*, 279–285. Amherst: Prometheus Books.

McCord, T., R. W. Carlson, P. D. Martin, et al. 1997. "Organics and other molecules in the surfaces of Callisto and Ganymede." *Science* 278:271–275.

McCoy, T. J. 1997. "A lively debate." *Nature* 386:557–558.

McKay, C. P. 1997. "The search for life on Mars." *Origins Life Evol. Biosphere* 27:263–289.

McKay, D. S., E. K. Gibson, Jr., K. L. Thomas-Keprta, H. Vali, C. S. Romaneck, S. J. Clemett, X. D. F. Chillier, C. R. Maechling, and R. N. Zare. 1996a. "Search for past life on Mars: Possible relic biogenic activity in Martian meteorite ALH84001." *Science* 273:924–930.

———, E. K. Gibson, Jr., and K. L. Thomas-Keprta. 1996b. "Past life on Mars? Response." *Science* 273:1640.

———, E. K. Gibson, Jr., K. L. Thomas-Keprta, and H. Vali. 1997. "No 'nanofossils' in Martian meteorite: Reply." *Nature* 390:455–456.

———, E. K. Gibson, Jr., and K. L. Thomas-Keprta. 1998a. "Earthly contaminants don't rule out Martian life." *Planetary Report* 18, no. 3:10–11.

———, S. W. Wenworth, K. L. Thopmas-Keprta, F. Westall, and E. K. Gibson, Jr. 1998b. "Possible bacteria in Nakhla." 30th Lunar and Planetary Science Conference. abstract 1816. see Web site http://cass.jsc.nasa.gov/epi/meteorites/30thlpscabs.html.

McKinnon, W. B. 1997. "Sighting the seas of Europa." *Nature* 386:765–767.

Mclaughlin, P. 1990. *Kant's Critique of Teleology in Biological Explanation*. Lampeter: Edwin Mellen Press.

McMullin, E. 1985. "Introduction: Evolution and creation." In E. McMullin, ed., *Evolution and Creation*. 1–56. Notre Dame, Ind.: University of Notre Dame Press.

Mehta, A., and G. Baker. 1991. "Self-organizing sand pile." *New Scientist*, 15 June, 40–43.

Michel, F., and E. Westhof. 1996. "Visualizing the logic behind RNA self-assembly." *Science* 273:1676–1677.

Miller, S. L. 1953. "A production of amino acids under possible primitive Earth conditions." *Science* 117:528–529.

———. 1987. "Which organic compounds could have occurred on the prebiotic Earth?" In Evolution of Catalytic Function. *Cold Spring Harbor Symposium. Quant. Biol.* 52:17–27.

———, and L. E. Orgel. 1973. *The Origins of Life*. Englewood Cliffs, N.J.: Prentice Hall.

———, and J. L. Bada. 1988. "Submarine hot springs and the origin of life." *Nature* 334:155–176.

Mills, D. R., R. L. Peterson, and S. Speigelman. 1967. "An extracellular Darwinian experiment with a self-duplicating nucleic acid molecule." *Proc. Natl. Acad. Sci. USA* 58:217.

Mittlefehldt, D. W. 1997. "The source of ALH84001." *Planetary Report* 17, no. 1:8–11.

Mojzsis, S. J., G. Arrhenius, K. D. McKeegan, T. M. Harrison, A. P. Nutman, and C. R. L. Friend. 1996. "Evidence for life on Earth before 3800 million years ago." *Nature* 384:55–59.

Monod, J. 1974. *Chance and Necessity*. Glasgow: Collins. Fontana Books.

Morowitz, H. J. 1992. *Beginnings of Cellular Life*. New Haven, Conn.: Yale University Press.

———, B. Heinz, and D. W. Deamer. 1988. "The chemical logic of a minimal protocell." *Origins Life Evol. Biosphere* 18:281–287.

Morrison, P. 1997. "Pattern in the rise of high technology." In C. B. Cosmovici, S. Bowyer, and D. Werthimer, eds., *Astronomical and Biochemical Origins and the Search for Life in the Universe*, 571–583. Proceedings of the 5th International Conference on Bioastronomy. Bologna, Italy: Editrice Compositori.

Morton, O. 1999. "To Mars, en masse." *Science* 283:1103–1104.

Muller, H. J. 1966. "The gene material as the initiator and the organizing basis of life." *American Naturalist* 100, No. 915:493–517.

Nealson, K. H. 1997. "Nannobacteria: Size limits and evidence." *Science* 276:1776.

Needham, J. T. 1748. "A summary of some late observations upon the generation, composition, and decomposition of animal and vegetable substances." *Phil. Trans. Royal Soc.* 45:615–666.

Nicolis, G., and G. Prigogine. 1977. *Self-Organization in Nonequilibrium Systems.* New York: John Wiley.

Nielsen, P. E. 1993. "Peptide nucleic acid (PNA): A model structure for the primordial genetic material?" *Origins Life Evol. Biosphere* 23:323–327.

Noller, H. F., V. Hoffarth, and L. Zimniak. 1992. "Unusual resistance of peptidyl transferase to protein extraction procedures." *Science* 256:1416–1419.

Norris, V., and D. J. Raine. 1998. "A fission-fusion origin for life." *Origins Life Evol. Biosphere* 28:523–537.

Nussbaum, M. C. 1978. "Aristotle on teleological explanations." In M. C. Nussbaum, *Aristotle's De Motu Animalium: Text with Translation, Commentary, and Interpretive Essays*, 59–106. Princeton, N.J.: Princeton University Press.

Oken, L. 1847. *Elements of Physico-philosophy.* Trans. A. Tulk. London: Ray Society.

Olby, R. C. 1966. *The Origins of Mendelism.* London: Constable.

Oparin, A. I. 1953 [1936]. *Origin of Life.* Trans. S. Morgulis. New York: Dover Publications.

———. 1961. *Life: Its Nature, Origin and Development.* New York: Academic Press.

———. 1965. "History of the subject matter of the conference." In S. W. Fox, ed., *The Origin of Prebiological Systems and of their Molecular Matrices*, 91–96. New York: Academic Press.

———. 1967. "The origin of life." Trans. A. Synge. In J. D. Bernal, *The Origin of Life*, 199–234. London: Weidenfeld and Nicolson. (Originally published in Moscow in 1924.)

———. 1968. *Genesis and Evolutionary Development of Life.* Trans. E. Maas. New York: Academic Press.

Orgel, L. E. 1968. "Evolution of the genetic apparatus." *J. Mol. Biol.* 38:381–393.

———. 1973. *The Origins of Life.* New York: John Wiley & Sons.

———. 1994. "The origin of life on the earth." *Scientific American*, October, 53–61.

———. 1998. "The origin of life—a review of facts and speculations." *TIBS* 23:491–495.

Oró, J. 1961. "Mechanism of synthesis of adenine from hydrogen cyanide under possible primitive earth conditions." *Nature* 191:1193–1194.

———, and C. B. Cosmovici. 1997. "Comets and life on the primitive Earth." In C. B. Cosmovici, S. Bowyer, and D. Werthimer, eds., *Astronomical and Biochemical Origins and the Search for Life in the Universe*, 97–120. Proceedings of the 5th International Conference on Bioastronomy. Bologna, Italy: Editrice Compositori.

Owen, T. 1980. "The search for early forms of life in other planetary systems: Future possibilities afforded by spectroscopic techniques." In M. D. Papagiannis, ed., *Strategies for the Search for Life in the Universe*, 177. Dordrecht, Netherlands: Reidel.

———. 1994. "The search for other planets: Clues from the solar system." *Astrophys. and Space Sci.* 212:1–11.

———. 1997. "What happened to Mars' ancient atmosphere?" *Planetary Report* 17, no. 1:28–29.

———, and A. Bar-Nun. 1995. "Comets, impacts and atmospheres." *Icarus* 116:215–226.

Pace, N. 1991. "Origin of life—facing up to the physical setting." *Cell* 65:531–533.

Pagels, H. R. 1988. *The Dreams of Reason: The Computer and the Rise of Sciences of Complexity.* New York: Simon and Schuster.

Paley, W. 1970 [1802]. *Natural Theology: or, Evidences of the Existence and Attributes of the Deity, Collected from the Appearances of Nature.* Farnborough, U.K.: Gregg.

Parsons, P. 1996. "Dusting off panspermia." *Nature* 383:221–222.

Pasteur, L. 1922–1939. *Ouevres.* Ed. Pasteur Vallery-Radot. 7 vols. Paris: Masson et Cie.

Pauling, L. 1959. *General Chemistry.* 2nd ed. San Francisco: W. H. Freeman.

Pendelton, Y. J., 1997. "Detection of organic matter in interstellar grains." *Origins Life Evol. Biosphere* 27:53–78.

Pennisi, E. 1998. "Genome data shake tree of life." *Science* 280:672–674.

Pennock, R. T. 1996. "Naturalism, evidence and creationism: The case of Phillip Johnson." *Biology & Philosophy* 11:543–559.

Piccirilli, J. A. 1995. "RNA seeks its maker." *Nature* 376:548–549.

———, T. S. McConnell, A. J. Zaug, H. F. Noller, and T. R. Cech. 1992. "Aminoacyl esterase activity of the *Tetrahymena* ribozyme." *Science* 256:1420–1424.

Planetary Society's Web site: http://planetary.org.

Podolsky, S. 1996. "The role of the virus in origin-of-life theorizing." *Journal of the History of Biology* 29:79–126.

Pohorille, A., and M. A. Wilson. 1995. "Molecular dynamics studies of simple membrane-water interfaces: Structure and function in the beginnings of cellular life." *Origins Life Evol. Biosphere* 25:21–46.

Ponnamperuma, C., and R. Navarro-Gonzales. 1995. "Primordial organic cosmochemistry." In B. Zuckerman and M. H. Hart, eds., *Extraterrestrials: Where Are They?* 2nd ed., 116–124. Cambridge: Cambridge University Press.

Popper, K. 1974. "Reduction and the incompleteness of science." In F. Ayala and T. Dobzhansky, eds., *Studies in the Philosophy of Biology.* Berkeley: University of California Press.

Portalie, E. 1960. *A Guide to the Thought of St. Augustine.* Trans. R. Bastian. Chicago: Regnery.

Postgate, J. 1994. *The Outer Reach of Life.* Cambridge: Cambridge University Press.

Pouchet, F. 1858. "Note sur les proto-organismes végétaux et animaux, nés spontanément dans l'air artificiel et dans le gaz oxygène." *Comptes rendus* 47:979–84.

———. 1859. *Hétérogénie, ou traité de la génération spontanée.* Paris: Baillière.

Psenner, R., and M. Loferer. 1997. "Nannobacteria: Size limits and evidence." *Science* 276:1776–1777.

Puget, J. L., and A. Léger. 1989. "A new component of the interstellar matter: Small grains and large aromatic molecules." *Ann. Rev. Astron. Astrophys.* 27:161–198.

Rasio, F. A., and E. B. Ford, 1996. "A dynamical formation process for planets in short-period orbits." *Science* 274:954.

Raulin-Cerceau, R., M. Maurel, and J. Schneider. 1998. "From panspermia to bioastronomy, the evolution of the hypothesis of universal life." *Origins Life Evol. Biosphere* 28:597–612.

Roll-Hansen, N. 1983. "The death of spontaneous generation and the birth of the gene: Two case studies of relativism." *Social Studies of Science* 13:481–519.

———. 1988. "The progress of eugenics: Growth of knowledge and change in ideology." *History of Science* 26:319–333.

Ruse, M. 1996. "Booknotes." *Biology & Philosophy* 11, no. 4:571–574.

Russell, M. J., A. J. Hall, A. G. Cairns-Smith, and P. S. Braterman. 1988. "Submarine hot springs and the origin of life." *Nature* 336:117.

Safhill, R., H. Schneider-Bernloehr, L. E. Orgel, and S. Speigelman. 1970. "*In vitro* selection of bacteriophage Qβ variants resistant to ethidium bromide." *J. Mol. Biol.* 51:531.

Sagan, C., ed., 1973. *Communication with Extraterrestrial Intelligence.* Cambridge, Mass.: MIT Press.

———. 1991. "Life." In *Encyclopaedia Britannica.* 15th ed., 22:985–1002.

————. 1994. "The search for extraterrestrial life." *Scientific American*, October, 71–75.

————. 1995. "Debate with E. Mayr." *Bioastronomy News* 7, no. 3.

————, R. W. Thompson, R. Carlson, D. Gurnett, and C. Hord. 1993. "A search for life on Earth from the Galileo spacecraft." *Nature* 365:715–721.

————, and S. Ostro. 1994. "Long-range consequences of interplanetary collisions." *Issues in Science and Technology* 10, no. 4:67–72.

————, and C. Chyba, 1997. "The early faint sun paradox: Organic shielding of ultraviolet labile greenhouse gases." *Science* 276:1217–1221.

Sakai, H., T. Gamo, E.-S. Kim, M. Tsutsumi, T. Tanaka, J. Ishibashi, H. Wakita, M. Yamano, and T. Oomori. 1990. "Venting of carbon dioxide–rich fluid and hydrate formation in mid-Okinawa Trough backarc basin." *Science* 248:1093–1096.

Sambursky, S. 1960. *The Physical World of the Greeks*. London: Routledge & Kegan Paul.

Schidlowski, M. 1988. "A 3,800–million-year isotopic record of life from carbon in sedimentary rocks." *Nature* 333:313–318.

Schoonen, M. A. A., Y. Xu, and J. Bebie. 1999. "Energetics and kinetics of the prebiotic synthesis of simple organic acids and amino acids with the FeS-H_2S/FeS_2 redox couple as reductant. *Origins Life Evol. Biosphere* 29:5–32.

Schopf, J. W, ed., 1983. *Earth's Earliest Biosphere: Its Origin and Evolution*. Princeton, N.J.: Princeton University Press.

————. 1993. "Microfossils of the early Archean apex chert: New evidence of the antiquity of life." *Science* 260:640–646.

Schrödinger, E. 1948. *What Is Life?* Cambridge, U.K.: Cambridge University Press.

Score, R. 1997. "Finding ALH84001." *Planetary Report* 17, no. 1:5–7.

Scott, E. C. 1997. "Creationists and the Pope's statement." *Quarterly Review of Biology* 72, no. 4:401–406.

Segré, D., D. Lancet, O. Kedem, and Y. Pilpel. 1998. "Graded autocatalysis replication domain (GARD): Kinetic analysis of self-replication in mutually catalytic sets." *Origins Life Evol. Biosphere* 28:501–514.

Shapiro, A. 1993. "Did Michael Ruse give away the store?" *National Center for Science Education Reports*, spring, 20–21.

Shapiro, R. 1986. *Origins: A Skeptic's Guide to the Creation of Life in the Universe*. New York: Summit Books.

————. 1988. "Prebiotic ribose synthesis: A critical analysis." *Origins Life Evol. Biosphere* 18:71–85.

————. 1995. "The prebiotic role of adenine: A critical analysis." *Origins Life Evol. Biosphere* 25:83–98.

————, and G. Feinberg. 1990. "Possible forms of life in environments very different from the Earth." In J. Leslie, ed., *Physical Cosmology and Philosophy*, 248–255. New York: Macmillan.

Shklovskii, I. S., and C. Sagan. 1966. *Intelligent Life in the Universe*. New York: Delta.

Shock, E. L. 1992a. "Chemical environments of submarine hydrothermal systems." *Origins Life Evol. Biosphere* 22:67–107.

————. 1992b. "Hydrothermal organic synthesis experiments." *Origins Life Evol. Biosphere* 22:135–146.

————, T. McCollom, and M. D. Schulte. 1995. "Geochemical constraints on chemolithoautotrophic reactions in hydrothermal systems." *Origins Life Evol. Biosphere* 25:141–159.

Sievers, D., and G. von Kiedrowski. 1994. "Self-replication of complementary nucleotide-based oligomers." *Nature* 369:221–224.

Simoneit, B. R. T. 1995. "Evidence for organic synthesis in high temperature aqueous media—facts and prognosis." *Origins Life Evol. Biosphere* 25:119–140.

Sleep, N. H., K. J. Zahnle, J. F. Kasting, and H. J. Morowitz. 1989. "Annihilation of ecosystems by large asteroid impacts on the early Earth." *Nature* 342:139–142.

Smolin, L. 1997. *The Life of the Cosmos*. New York: Oxford University Press.

Snyder, L. E. 1997. "The search for interstellar glycine." *Origins Life Evol. Biosphere* 27:115–133.

Spallanzani, L. 1803. *Tracts on the natural history of animals and vegetables*. Trans. J. G. Dalyell. Edinburgh.

Spiegelman, S. 1967. "An *in vitro* analysis of a replicating molecule." *American Scientist* 55:221–264.

Stein, D. L, ed., 1989. *Lectures in the Sciences of Complexity*. New York: Addison-Wesley.

Stryer, L. 1995. *Biochemistry*. 4th ed. New York: W. H. Freeman.

Sullivan, W. 1970. *We Are Not Alone*. Harmondsworth, U.K.: Penguin Books.

Tarter, J. 1995. "One attempt to find where they are: NASA's high resolution microwave survey." In B. Zuckerman, and M. H. Hart, eds., *Extraterrestrials: Where Are They?* 2nd ed., 9–19. Cambridge, U.K.: Cambridge University Press.

Taylor, R. L. 1997. "Life on another world: Can religious cope?" *Planetary Report* 17, no. 1:21.

Thaxton, C. B., W. L. Bradley, and R. L. Olsen. 1984. *The Mystery of Life's Origin*. New York: Philosophical Library.

Thomas Aquinas. 1945 [1265]. *Summa theologica*. In A. C. Pegis, ed., *Basic Writings of Thomas Aquinas*, vol. 1. New York: Random House.

Thomas-Keprta, K. L., D. A. Bazylinski, D. C. Golden, S. J. Wentworth, E. K. Gibson, Jr., and D. S. McKay. 1998. "Magnetite from ALH84001 carbonate globules: Evidence of biogenic signature?" In Lunar Planet. Sci. 29, abstract no. 1494. Lunar and Planetary Institute, Houston, Texas (CD-ROM). See *cass.jsc.nasa.gov/meetings/meetings.html*.

Thompson, W. R., G. D. McDonald, and C. Sagan. 1994. ""The Titan haze revisited: Magnetospheric energy sources and quantitative tholin yields." *Icarus* 112:376–381.

Thomson, W. 1871. Presidential Address to the British Association for the Advancement of Science, delivered in Edinburgh in 1871. London.

Tielens, A. G. G. M., and S. B. Charnley. 1997. "Circumstellar and interstellar synthesis of organic molecules." *Origins Life Evol. Biosphere* 27:23–51.

Toulmin, S., and J. Goodfield. 1982. *The Architecture of Matter*. Chicago: University of Chicago Press.

Troland, L. T. 1914. "The chemical origin and regulation of life." *Monist* 24:92–133.

———. 1916. "The enzyme theory of life." *Cleveland Medical Journal* 15, no. 6:377–385.

———. 1917. "Biological enigmas and the theory of enzyme action." *American Naturalist* 51, no. 606:321–350.

Urey, H. 1952a. *The Planets: Their Origin and Development*. New Haven: Yale University Press.

———. 1952b. "On the early chemical history of the earth and the origin of life." *Proc. Natl. Acad. Sci. USA* 38:351–363.

Vallery-Radot, R. 1926. *The Life of Pasteur*. Trans. Mrs. R. L. Devonshire. New York: Doubleday.

Vandervliet, G. 1971. *Microbiology and the Spontaneous Generation Debate during the 1870s*. Lawrence, Kans.: Coronado Press.

Vaughn, J. 1998. "The sun sets on Mars Pathfinder." *Planetary Report* 18, no. 1:4–5.

Veverka, J. 1998. "The Contour mission: seeking clues to our origin." *Planetary Report* 18, no. 2:12–15.

Virchow, R. 1971 [1858]. *Cellular Pathology*. Trans. F. Chance. New York: Dover.

Vogel, G. 1998a. "A sulfurous start for protein synthesis?" *Science* 281:627.

———. 1998b. "Finding life's limits." *Science* 282:1399.

Von Hippel, F., and T. von Hippel. 1996. "Past life on Mars?" *Science* 273:1639.

Von Kiedrowski, G. 1986. "A self-replicating hexadeoxynucleotide." *Angew. Chem. Int. Ed. Engl.* 25:10, 932–935.

———. 1996. "Primodial soup or crêpes?" *Nature* 381:20–21.

Von Neuman, J. 1961–1963[1948]. "The general and logical theory of automata." In A. H. Taub, ed., *J. Von Neuman, Collected Works*, 5:288–328. New York: Macmillan.

Wächtershäuser, G. 1992. "Groundwork for an evolutionary biochemistry: The iron-sulphur world." *Prog. Biophys. Molec. Biol.* 58:85–201.

Waldrop, M. M. 1990. " Spontaneous order, evolution, and life." *Science* 247:1543–1545.

Warren, P. H. 1998. "Petrologic evidence for low-temperature, possibly flood evaporitic origin of carbonates in the ALH84001 meteorite." *Journal Geophys. Res.* 103:16759–16773.

Watson, J. D., and F. H. Crick. 1953a. "Molecular structure of nucleic acids: A structure for DNA." *Nature* 171:737–738.

———, and F. H. Crick. 1953b. "Genetic implications of the structure of DNA." *Nature* 171:964–967.

Web sites of missions to detect Earth-like planets: *http://ast.star.rl.ac.uk:80/darwin/* and *www.kepler.arc.nasa.gov/*.

Weber, B. H., D. J. Depew, and J. D. Smith, eds., 1988. *Entropy, Information and Evolution: New Perspectives on Physical and Biological Evolution.* Cambridge, Mass.: MIT Press.

Weiner, A. M., and N. Maizels. 1991. "The genomic tag model for the origin of protein synthesis." In S. Osawa, and T. Honjo, eds., *Evolution of Life.* Tokyo: Springer-Verlag.

Wells, H. G., J. Huxley, and G. P. Wells. 1934. *The Science of Life.* New York: Literary Guild.

Wetherill, G. W. 1994. "Possible consequences of the absence of 'Jupiters' in planetary systems." *Astro. Phys. Space Sci.* 212:23–32.

Whittet, D. C. B. 1997. "Is extraterrestrial organic matter relevant to the origin of life on Earth?" *Origins Life Evol. Biosphere* 27:249–262.

Wicken, J. S. 1987. *Evolution, Thermodynamics and Information.* New York: Oxford University Press.

Williams, D. M., J. F. Kasting, and R. A. Wade, 1997. "Habitable moons around extrasolar giant planets." *Nature* 385:234–235.

Williams, N. 1996. "Genome projects: Yeast genome sequence ferments new research." *Science* 272:481.

Wilson, C., and J. W. Szostak. 1995. "*In vitro* evolution of a self-alkylating ribozyme." *Nature* 374:777–782.

Wittung, P., P. E. Nielsen, O. Buchardt, M. Egholm, and B. Nordén. 1994. "DNA-like double helix formed by peptide nucleic acid." *Nature* 368:561–563.

Woese, C. R. 1967. *The Genetic Code—the Molecular Basis for Genetic Expression.* New York: Harper and Row.

———. 1987. "Bacterial evolution." *Microbiol. Rev.* 51:221–271.

———. 1998. "The universal ancestor." *Proc. Natl. Acad. Sci. USA* 95:6854–6859.

———. and G. E. Fox. 1977. "Phylogenetic structure of the prokaryotic domain: The primary kingdoms." *Proc. Natl. Acad. Sci. USA* 74:5088–5090.

Wolszczan, A. 1994. "Confirmation of Earth-mass planets orbiting the millisecond pulsar PSR B1257 + 12." *Science* 264:538–542.

Wright, M. C., and G. F. Joyce. 1997. "Continuous *in vitro* evolution of catalytic functions." *Science* 276:614–617.

Zahnle, K. J., and N. H. Sleep. 1996. "Impacts and the early evolution of life." In P. J. Thomas, C. F. Chyba, and C. P. McKay, eds., *Comets and the Origin and Evolution of Life,* 175–208. New York: Springer-Verlag.

Zielinski, W. S., and L. E. Orgel. 1987. "Oligoaminonucleoside phosphoramidates. Oligomerization of dimers of 3'-amino-3'-deoxy-nucleotides (GC and CG) in aqueous solution." *Nucleic Acids Res.* 15:4, 1699–1715.

Zuckerman, B. 1995. "Preface to the first edition. 1980." In B. Zuckerman and M. H. Hart, eds., *Extraterrestrials: Where Are They?* 2nd ed., xi–xii. Cambridge: Cambridge University Press.

———, and M. H. Hart, eds. 1995. *Extraterrestrials: Where Are They?* 2nd ed. Cambridge: Cambridge University Press.

Index

abiogenesis, 48; blow to, following discrediting of *Bathybius haeckelii*, 79; Pflüger's version of, 58; 19th-century Darwinists and, 57, 58, 179; Pasteur-Pouchet conflict and, 50–53; Pouchet's rejection of, 48; a separate scientific question from heterogenesis? 50–52

accretion model of the formation of planets, 114, 261, 262; Apollo space program and, 114; "cold homogeneous accretion model", 114; delivery to Earth of organic compounds and water and, 115; and differentiation between inner and outer planets, 262; "hot heterogeneous accretion model", 114; Oparin-Haldane scenario and, 115; planetesimals and, 114, 261, 262; protoplanetary disks and, 261–262; reevaluation of, after discovery of extrasolar planets, 262–264; Schmidt, Otto, and, 114

Altman, Sidney, 138

Anaximander. *See* matter, first theories of

Anaximenes. *See* matter, first theories of

Anders, Edward, 230

Anderson, David, 272

Angel, Robert, 266

animalcules, 24, 27. *See also* spontaneous generation

animists, 22: awareness of biological complexity, 23; reaction to extreme mechanism, 22, 24; religious connotations of soul, 23

antiquity of life on Earth, evidence for: biologically derived material from rocks aged 3.85 billion years, 125; fossils of bacteria cells aged 2.7 billion years, 124; metamorphism in older rocks and, 129; new techniques of measuring carbon-isotope composition and, 124; panspermia and, 234–235; ratio of two carbon isotopes in organic material and, 124–125; root of life beyond the Earth? 120, 134; Schopf's findings of fossils aged 3.5 billion years, 124, 227; stromatolites and, 124. *See also* "time window" for the origin of life

arch-and-scaffolding metaphor, 185–193; application of, to evolutionary developments, 204; Cairns-Smith's clay scaffold, 185–186; Cairns-Smith's metaphor as a unifying theme of all origin-of-life theories, 186–187, 242; Cairns-Smith's

About the Author

Iris Fry teaches the history and philosophy of biology at the Cohn Institute for the History and Philosophy of Science and Ideas at Tel Aviv University and in the department of humanities and arts at the Technion–Israel Institute of Technology.

Initially trained in chemistry and biochemistry at the Hebrew University, Jerusalem, she subsequently studied philosophy and the history and philosophy of science at Haifa and Tel Aviv Universities. Her publications deal with the concept of teleology in the philosophy of Immanuel Kant, the history of evolutionary ideas and the origin of life. Her book, *The Origin of Life–Mystery or Scientific Problem?* was published in Israel in 1997.

Printed in the United States
112432LV00003B/91-105/A